T0295296

Data Integration, Manipulation and Visualization of Phylogenetic Trees

Data Integration, Manipulation and Visualization of Phylogenetic Trees introduces and demonstrates data integration, manipulation and visualization of phylogenetic trees using a suite of R packages, tidytree, treeio, ggtree and ggtreeExtra. Using the most comprehensive packages for phylogenetic data integration and visualization, contains numerous examples that can be used for teaching and learning. Ideal for undergraduate readers and researchers with a working knowledge of R and ggplot2.

Key Features:

- Manipulating phylogenetic tree with associated data using tidy verbs
- Integrating phylogenetic data from diverse sources
- Visualizing phylogenetic data using grammar of graphics

Guangchuang Yu (https://yulab-smu.top) is a professor of bioinformatics, director of the Department of Bioinformatics, Southern Medical University. He earned his Ph.D. from the School of Public Health, The University of Hong Kong. As an active R user, he has authored several R packages and has supervised post-graduate students to develop a few other packages, including ggmsa, ggtreeExtra, MicrobiomeProfiler and MicrobiotaProcess.

Chapman & Hall/CRC Computational Biology Series

About the Series:
This series aims to capture new developments in computational biology, as well as high-quality work summarizing or contributing to more established topics. Publishing a broad range of reference works, textbooks, and handbooks, the series is designed to appeal to students, researchers, and professionals in all areas of computational biology, including genomics, proteomics, and cancer computational biology, as well as interdisciplinary researchers involved in associated fields, such as bioinformatics and systems biology.

Introduction to Bioinformatics with R: A Practical Guide for Biologists
Edward Curry

Analyzing High-Dimensional Gene Expression and DNA Methylation Data with R
Hongmei Zhang

Introduction to Computational Proteomics
Golan Yona

Glycome Informatics: Methods and Applications
Kiyoko F. Aoki-Kinoshita

Computational Biology: A Statistical Mechanics Perspective
Ralf Blossey

Computational Hydrodynamics of Capsules and Biological Cells
Constantine Pozrikidis

Computational Systems Biology Approaches in Cancer Research
Inna Kuperstein, Emmanuel Barillot

Clustering in Bioinformatics and Drug Discovery
John David MacCuish, Norah E. MacCuish

Metabolomics: Practical Guide to Design and Analysis
Ron Wehrens, Reza Salek

An Introduction to Systems Biology: Design Principles of Biological Circuits
2nd Edition
Uri Alon

Computational Biology: A Statistical Mechanics Perspective
Second Edition
Ralf Blossey

Stochastic Modelling for Systems Biology
Third Edition
Darren J. Wilkinson

Computational Genomics with R
Altuna Akalin, Bora Uyar, Vedran Franke, Jonathan Ronen

An Introduction to Computational Systems Biology: Systems-level Modelling of Cellular Networks
Karthik Raman

Virus Bioinformatics
Dmitrij Frishman, Manuela Marz

Multivariate Data Integration Using R: Methods and Applications with the mixOmics Package
Kim-Anh LeCao, Zoe Marie Welham

Bioinformatics
A Practical Guide to NCBI Databases and Sequence Alignments
Hamid D. Ismail

Data Integration, Manipulation and Visualization of Phylogenetic Trees
Guangchuang Yu

For more information about this series please visit:
https://www.routledge.com/Chapman--HallCRC-Computational-Biology-Series/book-series/CRCCBS

Data Integration, Manipulation and Visualization of Phylogenetic Trees

Guangchuang Yu

CRC Press is an imprint of the
Taylor & Francis Group, an **informa** business
A CHAPMAN & HALL BOOK

First edition published 2023
by CRC Press
6000 Broken Sound Parkway NW, Suite 300, Boca Raton, FL 33487-2742

and by CRC Press
4 Park Square, Milton Park, Abingdon, Oxon, OX14 4RN

CRC Press is an imprint of Taylor & Francis Group, LLC

Library of Congress Cataloging-in-Publication Data

Names: Yu, Guangchuang, author.
Title: Data integration, manipulation and visualization of phylogenetic trees / Guangchuang Yu.
Description: First edition. | Boca Raton : CRC Press, [2022] | Series: Chapman & Hall/CRC computational biology series | Includes bibliographical references and index.
Identifiers: LCCN 2022001948 (print) | LCCN 2022001949 (ebook) | ISBN 9781032233574 (hbk) | ISBN 9781032245546 (pbk) | ISBN 9781003279242 (ebk)
Subjects: LCSH: Phylogeny--Data processing.
Classification: LCC QH367.5 .Y84 2022 (print) | LCC QH367.5 (ebook) | DDC 576.8/8--dc23/eng/20220314
LC record available at https://lccn.loc.gov/2022001948
LC ebook record available at https://lccn.loc.gov/2022001949

ISBN: 978-1-032-23357-4 (hbk)
ISBN: 978-1-032-24554-6 (pbk)
ISBN: 978-1-003-27924-2 (ebk)

DOI: 10.1201/9781003279242

Typeset in CMR10 font
by KnowledgeWorks Global Ltd.

Publisher's note: This book has been prepared from camera-ready copy provided by the authors.

Contents

List of Figures

List of Tables

Preface

I am so excited to have this book published. The book is meant as a guide for data integration, manipulation and visualization of phylogenetic trees using a suite of R packages, **tidytree**, **treeio**, **ggtree** and **ggtreeExtra**. Hence, if you are starting to read this book, we assume you have a working knowledge of how to use R and **ggplot2**.

The development of the **ggtree** package started during my PhD study at the University of Hong Kong. I joined the State Key Laboratory of Emerging Infectious Diseases (SKLEID) under the supervision of Yi Guan and Tommy Lam. I was asked to provide assistance to modify the newick tree string to incorporate some additional information, such as amino acid substitutions, in the internal node labels of the phylogeny for visualization. I wrote an R script to do it and soon realized that most phylogenetic tree visualization software can only display one type of data through node labels. Basically, we cannot display two data variables at the same time for comparative analysis. In order to produce tree graphs displaying different types of branch/node associated information, such as bootstrap values and substitutions, people mostly relied on post-processing image software. This situation motivates me to develop **ggtree**. First of all, I think a good user interface must fully support the **ggplot2** syntax, which allows us to draw graphs by superimposing layers. In this way, simple graphs are simple, and complex graphs are just a combination of simple layers, which are easy to generate.

After several years of development, **ggtree** has evolved into a package suite, including **tidytree** for manipulating tree with data using the tidy interface; **treeio** for importing and exporting tree with richly annotated data; **ggtree** for tree visualization and annotation and **ggtreeExtra** for presenting data with a phylogeny side-by-side for a rectangular layout or in outer rings for a circular layout. The **ggtree** is a general tool that supports different types of tree and tree-like structures and can be applied to different disciplines to help researchers presenting and interpreting data in the evolutionary or hierarchical context.

Structure of the book

- Part I (Tree data input, output and manipulation) describes **treeio** package for tree data input and output, and **tidytree** package for tree data manipulation.

- Part II (Tree data visualization and annotation) introduces tree visualization and annotation using the grammar of graphic syntax implemented in the **ggtree** package. It emphasizes presenting tree-associated data on the tree.
- Part III (ggtree extensions) introduces **ggtreeExtra** for presenting data on circular layout trees and other extensions including **MicrobiotaProcess** and **tanggle** *etc.*
- Part IV (Miscellaneous topics) describes utilities provided by the **ggtree** package suite and presents a set of reproducible examples.

Software information and conventions

The R and core packages information when compiling this book is as follows:

```
R.version.string
```

```
## [1] "R version 4.1.2 (2021-11-01)"
```

```
library(treedataverse)
```

```
##  Attaching packages  treedataverse 0.0.1

##  ape          5.5            treeio      1.18.1
##  dplyr        1.0.7          ggtree      3.2.1
##  ggplot2      3.3.5          ggtreeExtra 1.4.1
##  tidytree     0.3.6
```

The **treedataverse** is a meta package to make it easy to install and load core packages for processing and visualizing tree with data using the packages described in this book. The installation guide for **treedataverse** can be found in FAQ.

The datasets used in this book have three sources:

1. Simulation data
2. Datasets in the R packages
3. Data downloaded from the Internet

In order to make the data downloaded from the Internet more accessible, we packed the data in an R package, **TDbook**, with detailed documentation of the original source, including URL, authors, and citation if the information is available. The **TDbook** is available on CRAN and can be installed using `install.packages("TDbook")`.

Package names in this book are formatted as bold text (*e.g.*, **ggtree**), and function names are followed by parentheses (*e.g.*, `treeio::read.beast()`). The double-colon operator (`::`) means accessing an object from a package.

Acknowledgments

Many people have contributed to this book with spelling and grammar corrections. I'd particularly like to thank Shuangbin Xu, Lin Li and Xiao Luo for their detailed technical reviews of the book, and Tiao You for designing the front cover of the book.

Many others have contributed during the development of the **ggtree** package suite. I would like to thank Hadley Wickham, for creating the **ggplot2** package that **ggtree** relies on; Tommy Tsan-Yuk Lam and Yi Guan for being great advisors and supporting the development of **ggtree** during my PhD; Richard Ree for inviting me to catalysis meeting on phylogenetic tree visualization; William Pearson for inviting me to publish a protocol paper of **ggtree** in the *Current Procotols in Bioinformatics* journal; Shuangbin Xu, Yonghe Xia, Justin Silverman, Bradley Jones, Watal M. Iwasaki, Ruizhu Huang, Casey Dunn, Tyler Bradley, Konstantinos Geles, Zebulun Arendsee and many others who have contributed source code or given me feedback; and last, but not least, the members of the **ggtree** mailing list[1], for providing many challenging problems that have helped improve the **ggtree** package suite.

[1]https://groups.google.com/forum/#!forum/bioc-ggtree

About the Author

Guangchuang Yu (https://yulab-smu.top) is a professor of Bioinformatics and director of the Department of Bioinformatics at Southern Medical University. He earned his Ph.D. from the School of Public Health, The University of Hong Kong. As an active R user, he has authored several R packages, such as **aplot**, **badger**, **ChIPseeker**, **clusterProfiler**, **DOSE**, **emojifont**, **enrichplot**, **ggbreak**, **ggfun**, **ggimage**, **ggplotify**, **ggtree**, **GOSemSim**, **hexSticker**, **meme**, **meshes**, **nCov2019**, **plotbb**, **ReactomePA**, **scatterpie**, **seqmagick**, **seqcombo**, **shadowtext**, **tidytree** and **treeio**. He has supervised post-graduate students to develop a few other packages, including **ggmsa**, **ggtreeExtra**, **MicrobiomeProfiler** and **MicrobiotaProcess**.

His research group aims to generate new insights into human health and disease through the development of new software tools and novel analysis of biomedical data. The software package developed by his research group helps biologists analyze data and reveal biological clues hidden in the data.

He has published several journal articles, including 5 highly cited papers (Yu et al., 2017, 2012; Yu, Wang, Yan, et al., 2015; Yu, Wang, & He, 2015; Yu & He, 2016). The articles have been cited more than 10,000 times. The ggtree (Yu et al., 2017) paper was selected as a feature article to celebrate the 10th anniversary of the launch of *Methods in Ecology and Evolution*[2]. He was one of the 2020 Highly Cited Chinese Researchers (Elsevier-Scopus) in Biomedical Engineering.

[2] 10th Anniversary Volume 8: Phylogenetic tree visualization with multivariate data: https://methodsblog.com/2020/11/19/ggtree-tree-visualization/

Part I: Tree data input, output, and manipulation

Chapter 1

Importing Tree with Data

1.1 Overview of Phylogenetic Tree Construction

Phylogenetic trees are used to describe genealogical relationships among a group of organisms, which can be constructed based on the genetic sequences of the organisms. A rooted phylogenetic tree represents a model of evolutionary history depicted by ancestor-descendant relationships between tree nodes and clustering of sister' orcousin' organisms at a different level of relatedness, as illustrated in Figure 1.1. In infectious disease research, phylogenetic trees are usually built from the pathogens' gene or genome sequences to show which pathogen sample is genetically closer to another sample, providing insights into the underlying unobserved epidemiologic linkage and the potential source of an outbreak.

A phylogenetic tree can be constructed from genetic sequences using distance-based methods or character-based methods. Distance-based methods, including the unweighted pair group method with arithmetic means (UPGMA) and the Neighbor-joining (NJ), are based on the matrix of pairwise genetic distances calculated between sequences. The character-based methods, including maximum parsimony (MP) (Fitch, 1971), maximum likelihood (ML) (Felsenstein, 1981), and Bayesian Markov Chain Monte Carlo (BMCMC) method (Rannala & Yang, 1996), are based on a mathematical model that describes the evolution of genetic characters and searches for the best phylogenetic tree according to their optimality criteria.

The MP method assumes that the evolutionary change is rare and minimizes the amount of character-state changes (*e.g.*, number of DNA substitutions). The criterion is similar to Occam's razor, that the simplest hypothesis that can explain the data is the best hypothesis. Unweighted parsimony assumes mutations across different characters (nucleotides or amino acids) are equally likely, while the weighted method assumes the unequal likelihood of mutations (*e.g.*, the third codon position is more liable than other codon positions; and the transition mutations have a higher frequency than transversion). The concept of the MP method is straightforward and intuitive, which is a probable reason for its popularity amongst biologists who

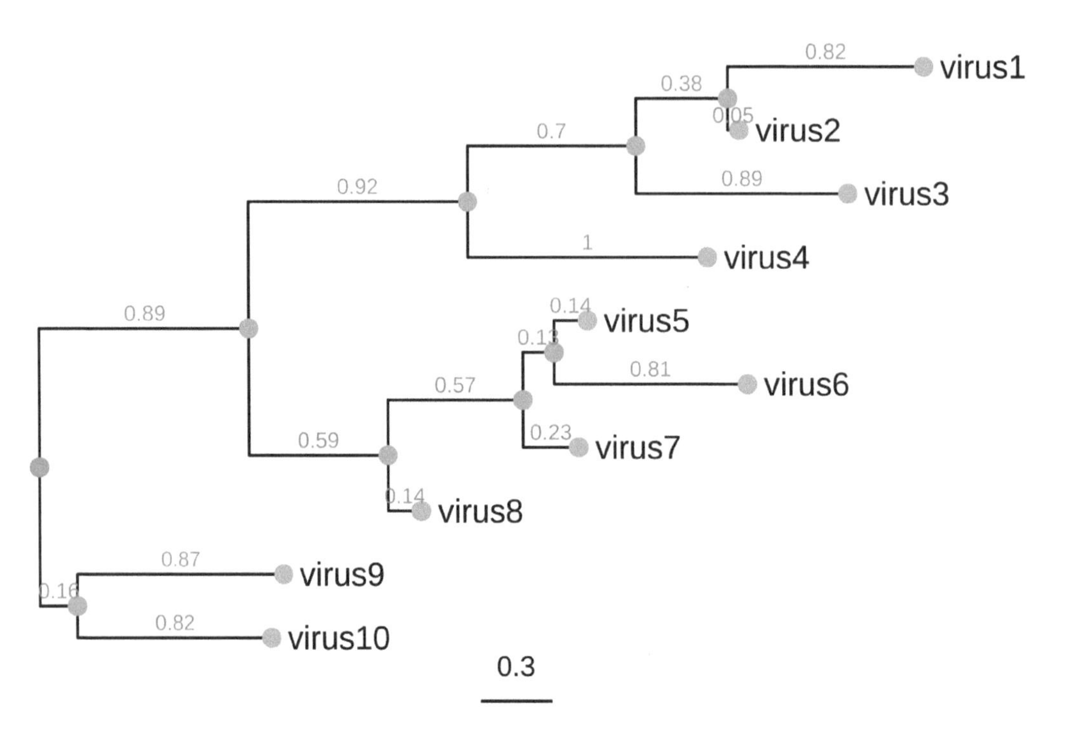

Figure 1.1: **Components of a phylogenetic tree.** External nodes (green circles), also called tips', represent actual organisms sampled and sequenced (*e.g.*, viruses in infectious disease research). They are thetaxa' in the terminology of evolutionary biology. The internal nodes (blue circles) represent hypothetical ancestors for the tips. The root (red circle) is the common ancestor of all species in the tree. The horizontal lines are branches and represent evolutionary changes (gray number) measured in a unit of time or genetic divergence. The bar at the bottom provides the scale of these branch lengths.

care more about the research question rather than the computational details of the analysis. However, this method has several disadvantages, in particular, the tree inference can be biased by the well-known systematic error called long-branch attraction (LBA) that incorrectly infer distantly related lineages as closely related (Felsenstein, 1978). This is because the MP method poorly takes into consideration of many sequence evolution factors (*e.g.*, reversals and convergence) that are hardly observable from the existing genetic data.

The maximum likelihood (ML) method and Bayesian Markov Chain Monte Carlo (BMCMC) method are the two most commonly used methods in phylogenetic tree construction and are most often used in scientific publications. ML and BMCMC methods require a substitution model of sequence evolution. Different sequence data have different substitution models to formulate the evolutionary process of DNA, codon and amino acid. There are several models for nucleotide substitution, including JC69, K2P, F81, HKY, and GTR (Arenas, 2015). These models can be used in conjunction with the rate variation across sites (denoted as $+\Gamma$)) (Yang, 1994) and the proportion of invariable sites (denoted as +I) (Shoemaker & Fitch, 1989).

Previous research (Lemmon & Moriarty, 2004) had suggested that misspecification of substitution model might bias phylogenetic inference. Procedural testing for the best-fit substitution model is recommended.

The optimal criterion of the ML method is to find the tree that maximizes the likelihood given the sequence data. The procedure of the ML method is simple: calculating the likelihood of a tree and optimizing its topology and branches (and the substitution model parameters, if not fixed) until the best tree is found. Heuristic search, such as those implemented in **PhyML** and **RAxML**, is often used to find the best tree based on the likelihood criterion. The Bayesian method finds the tree that maximizes posterior probability by sampling trees through MCMC based on the given substitution model. One of the advantages of BMCMC is that parameter variance and tree topological uncertainty, included by the posterior clade probability, can be naturally and conveniently obtained from the sampling trees in the MCMC process. Moreover, the influence of topological uncertainty on other parameter estimates is also naturally integrated into the BMCMC phylogenetic framework.

In a simple phylogenetic tree, data associated with the tree branches/nodes could be the branch lengths (indicating genetic or time divergence) and lineage supports such as bootstrap values estimated from bootstrapping procedure or posterior clade probability summarized from the sampled trees in the BMCMC analysis.

1.2 Phylogenetic Tree Formats

There are several file formats designed to store phylogenetic trees and the data associated with the nodes and branches. The three commonly used formats are Newick[1], NEXUS (Maddison et al., 1997), and Phylip (Felsenstein, 1989). Some formats (*e.g.*, NHX) are extended from the Newick format. Newick and NEXUS formats are supported as input by most of the software in evolutionary biology, while some of the software tools output newer standard files (*e.g.*, **BEAST** and **MrBayes**) by introducing new rules/data blocks for storing evolutionary inferences. In the other cases (*e.g.*, **PAML** and **r8s**), output log files are only recognized by their own single software.

1.2.1 Newick tree format

The Newick tree format is the standard for representing trees in computer-readable form.

The rooted tree shown in Figure 1.2 can be represented by the following sequence of characters as a Newick tree text.

```
((t2:0.04,t1:0.34):0.89,(t5:0.37,(t4:0.03,t3:0.67):0.9):0.59);
```

The tree text ends with a semicolon. Internal nodes are represented by a pair of matched parentheses. Between the parentheses are descendant nodes of that

[1]http://evolution.genetics.washington.edu/phylip/newick_doc.html

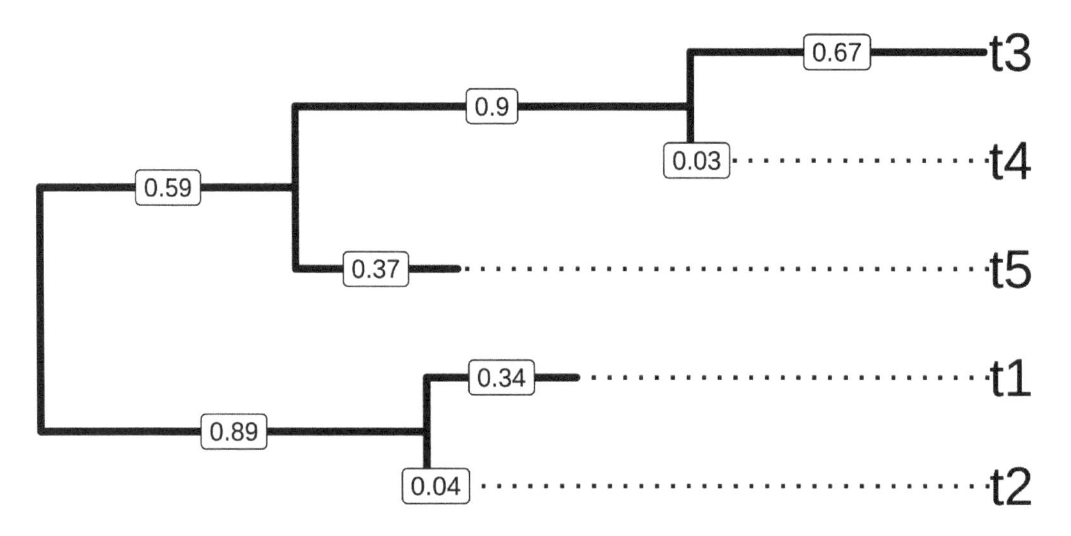

Figure 1.2: **A sample tree for demonstrating Newick text to encode tree structure.** Tips were aligned to the right-hand side and branch lengths were labeled on the middle of each branch.

node. For instance (t2:0.04,t1:0.34) represents the parent node of t2 and t1 that are the immediate descendants. Sibling nodes are separated by a comma and tips are represented by their names. A branch length (from the parent node to child node) is represented by a real number after the child node and is preceded by a colon. Singular data (*e.g.*, bootstrap values) associated with internal nodes or branches may be encoded as node labels and represented by the simple text/numbers before the colon.

Newick tree format was developed by Meacham in 1984 for the Phylogeny Inference Package or **PHYLIP** (Retief, 2000) package. Newick format is now the most widely used tree format and used by **PHYLIP**, **PAUP*** (Wilgenbusch & Swofford, 2003), *TREE-PUZZLE* (Schmidt et al., 2002), **MrBayes**, and many other applications. Phylip tree format contains Phylip multiple sequence alignment (MSA) with a corresponding Newick tree text that was built based on the MSA sequences in the same file.

1.2.2 NEXUS tree format

The NEXUS format (Maddison et al., 1997) incorporates Newick tree text with related information organized into separated units known as **blocks**. A NEXUS block has the following structure:

```
#NEXUS
...
BEGIN characters;
...
END;
```

For example, the above example tree can be saved as the following NEXUS format:

```
#NEXUS
[R-package APE, Wed Apr 20 10:10:41 2022]

BEGIN TAXA;
    DIMENSIONS NTAX = 5;
    TAXLABELS
        t2
        t1
        t5
        t4
        t3
    ;
END;
BEGIN TREES;
    TRANSLATE
        1   t2,
        2   t1,
        3   t5,
        4   t4,
        5   t3
    ;
    TREE * UNTITLED = [&R]
((1:0.04,2:0.34):0.89,(3:0.37,(4:0.03,5:0.67):0.9):0.59);
END;
```

Comments can be placed using square brackets. Some blocks can be recognized by most of the programs including `TAXA` (contains information of taxa), `DATA` (contains data matrix, *e.g.*, sequence alignment), and `TREE` (contains a phylogenetic tree, *i.e.*, Newick tree text). Notably, blocks can be very diverse and some of them are only recognized by one particular program. For example NEXUS file exported by **PAUP*** has a *paup* block that contains *PAUP** commands, whereas **FigTree** exports the NEXUS file with a *figtree* block that contains visualization settings. NEXUS organizes different types of data into different blocks, whereas programs that support reading NEXUS can parse some blocks they recognized and ignore those they could not. This is a good mechanism to allow different programs to use the same format without crashing when unsupported types of data are present. Notably, most of the programs only support parsing `TAXA`, `DATA`, and `TREE` blocks; therefore, a program/platform that could generically read all kinds of data blocks from the NEXUS would be useful for phylogenetic data integration.

The `DATA` block is widely used to store sequence alignment. For this purpose, the user can store tree and sequence data in Phylip format which are essentially Phylip multiple sequence alignment and Newick tree text, respectively. It is used in **PHYLIP**.

1.2.3 New Hampshire eXtended format

Newick, NEXUS, and phylip are mainly designed to store phylogenetic trees and basic singular data associated with internal nodes or branches. In addition to the singular data annotation at branches and nodes (mentioned above), New Hampshire eXtended (NHX) format, which is based on Newick (also called New Hampshire), was developed to introduce tags to associate multiple data fields with the tree nodes (both internal nodes and tips). Tags are placed after branch length and must be wrapped between `[&&NHX` and `]` which makes it possible to be compatible with NEXUS format as it defined characters between `[` and `]` as comments. NHX is also the output format of **PHYLDOG** (Boussau et al., 2013) and **RevBayes** (Höhna et al., 2016). A Tree Viewer (**ATV**) (Zmasek & Eddy, 2001) is a java tool that supports displaying annotation data stored in NHX format, but this package is no longer maintained.

Here is a sample tree from NHX definition document[2]:

```
(((ADH2:0.1[&&NHX:S=human], ADH1:0.11[&&NHX:S=human]):0.05
[&&NHX:S=primates:D=Y:B=100],ADHY:0.1[&&NHX:S=nematode],
ADHX:0.12[&&NHX:S=insect]):0.1[&&NHX:S=metazoa:D=N],(ADH4:0.09
[&&NHX:S=yeast],ADH3:0.13[&&NHX:S=yeast],ADH2:0.12[&&NHX:S=yeast],
ADH1:0.11[&&NHX:S=yeast]):0.1[&&NHX:S=Fungi])[&&NHX:D=N];
```

1.2.4 Jplace format

To store the Next Generation Sequencing (NGS) short reads mapped onto a phylogenetic tree (for metagenomic classification), Matsen (Matsen et al., 2012) proposed jplace format for such phylogenetic placements. Jplace format is based on JSON and contains four keys: `tree`, `fields`, `placements`, `metadata`, and `version`. The `tree` value contains tree text extended from Newick tree format by putting the edge label in brackets (if available) after branch length and putting the edge number in curly braces after the edge label. The `fields` value contains header information of placement data. The value of `placements` is a list of `pqueries`. Each `pquery` contains two keys: `p` for placements and `n` for name or `nm` for names with multiplicity. The value of `p` is a list of placement for `pqueries`.

Here is a jplace sample file:

```
{
    "tree": "(((((((A:4{1},B:4{2}):6{3},C:5{4}):8{5},D:6{6}):
    3{7},E:21{8}):10{9},((F:4{10},G:12{11}):14{12},H:8{13}):
    13{14}):13{15},((I:5{16},J:2{17}):30{18},(K:11{19},
    L:11{20}):2{21}):17{22}):4{23},M:56{24});",
    "placements": [
    {"p":[24, -61371.300778, 0.333344, 0.000003, 0.003887],
     "n":["AA"]
    },
```

[2]http://www.genetics.wustl.edu/eddy/forester/NHX.html

```
    {"p":[[1, -61312.210786, 0.333335, 0.000001, 0.000003],
          [2, -61322.210823, 0.333322, 0.000003, 0.000003],
          [3, -61352.210823, 0.333322, 0.000961, 0.000003]],
     "n":["BB"]
    },
    {"p":[[8, -61312.229128, 0.200011, 0.000001, 0.000003],
          [9, -61322.229179, 0.200000, 0.000003, 0.000003],
          [10, -61342.229223, 0.199992, 0.000003, 0.000003]],
    "n":["CC"]
    }
    ],
    "metadata": {"info": "a jplace sample file"},
    "version" : 2,
    "fields": ["edge_num", "likelihood", "like_weight_ratio",
    "distal_length", "pendant_length"
    ]
}
```


Jplace is the output format of **PPLACER** (Matsen et al., 2010) and Evolutionary Placement Algorithm (**EPA**) (Berger et al., 2011). But these two programs do not contain tools to visualize placement results. **PPLACER** provides `placeviz` to convert jplace file to phyloXML or Newick formats which can be visualized by **Archaeopteryx**.

1.2.5 Software outputs

RAxML (Stamatakis, 2014) can output Newick format by storing the bootstrap values as internal node labels. Another way that **RAxML** supports is to place bootstrap value inside square brackets and after branch length. This could not be supported by most of the software that supports Newick format where square brackets will be ignored.

BEAST (Bouckaert et al., 2014) output is based on NEXUS, and it also introduces square brackets in the tree block to store evolutionary evidence inferred by **BEAST**. Inside brackets, curly braces may also be incorporated if feature values have a length of more than 1 (*e.g.*, Highest Probability Density (HPD) or range of substitution rate). These brackets are placed between node and branch length (*i.e.*, after label if exists and before the colon). The bracket is not defined in Newick format and is a reserved character for NEXUS comment. So this information will be ignored for standard NEXUS parsers.

Here is a sample `TREE` block of the **BEAST** output:

```
TREE * TREE1 = [&R] (((11[&length=9.47]:9.39,14[&length=6.47]:6.39)
[&length=25.72]:25.44,4[&length=9.14]:8.82)[&length=3.01]:3.1,
(12[&length=0.62]:0.57,(10[&length=1.6]:1.56,(7[&length=5.21]:5.19,
((((2[&length=3.3]:3.26,(1[&length=1.34]:1.32,(6[&length=0.85]:0.83,
```

```
13[&length=0.85]:0.83)[&length=2.5]:2.49)[&length=0.97]:0.94)
[&length=0.5]:0.5,9[&length=1.76]:1.76)[&length=2.41]:2.36,
8[&length=2.19]:2.11)[&length=0.27]:0.24,(3[&length=3.33]:3.31,
(15[&length=5.29]:5.27,5[&length=3.29]:3.27)[&length=1.04]:1.04)
[&length=1.98]:2.04)[&length=2.83]:2.84)[&length=5.39]:5.37)
[&length=2.02]:2)[&length=4.35]:4.36)[&length=0];
```

BEAST output can contain many different evolutionary inferences, depending on the analysis models defined in *BEAUTi* for running. For example in molecular clock analysis, it contains `rate`, `length`, `height`, `posterior` and corresponding HPD and range for uncertainty estimation. `Rate` is the estimated evolutionary rate of the branch. `Length` is the length of the branch in years. `Height` is the time from node to root, while `posterior` is the Bayesian clade credibility value. The above example is the output tree of a molecular clock analysis and should contain these inferences. To save space, only the `length` estimation was shown above. Besides, Molecular Evolutionary Genetics Analysis (**MEGA**) (Kumar et al., 2016) also supports exporting trees in **BEAST** compatible Nexus format (see session 1.3.2).

MrBayes (Huelsenbeck & Ronquist, 2001) is a program that uses the Markov Chain Monte Carlo method to sample from the posterior probability distributions. Its output file annotates nodes and branches separately by two sets of square brackets. For example below, posterior clade probabilities for the node and branch length estimates for the branch:

```
  tree con_all_compat = [&U] (8[&prob=1.0]:2.94e-1[&length_mean=2.9e-1],
10[&prob=1.0]:2.25e-1[&length_mean=2.2e-1],((((1[&prob=1.0]:1.43e-1
[&length_mean=1.4e-1],2[&prob=1.0]:1.92e-1[&length_mean=1.9e-1])[&prob=1.0]:
1.24e-1[&length_mean=1.2e-1],9[&prob=1.0]:2.27e-1[&length_mean=2.2e-1])
[&prob=1.0]:1.72e-1[&length_mean=1.7e-1],12[&prob=1.0]:5.11e-1
[&length_mean=5.1e-1])[&prob=1.0]:1.76e-1[&length_mean=1.7e-1],
(((3[&prob=1.0]:5.46e-2[&length_mean=5.4e-2],(6[&prob=1.0]:1.03e-2
[&length_mean=1.0e-2],7[&prob=1.0]:7.13e-3[&length_mean=7.2e-3])[&prob=1.0]:
6.93e-2[&length_mean=6.9e-2])[&prob=1.0]:6.03e-2[&length_mean=6.0e-2],
(4[&prob=1.0]:6.27e-2[&length_mean=6.2e-2],5[&prob=1.0]:6.31e-2
[&length_mean=6.3e-2])[&prob=1.0]:6.07e-2[&length_mean=6.0e-2])[&prob=1.0]:,
1.80e-1[&length_mean=1.8e-1]11[&prob=1.0]:2.37e-1[&length_mean=2.3e-1])
[&prob=1.0]:4.05e-1[&length_mean=4.0e-1])[&prob=1.0]:1.16e+000
[&length_mean=1.162699558201079e+000])[&prob=1.0][&length_mean=0];
```

To save space, most of the inferences were removed and only contains `prob` for clade probability and `length_mean` for mean value of branch length. The full version of this file also contains `prob_stddev`, `prob_range`, `prob(percent)`, `prob+-sd` for probability inferences and `length_median`, `length_95%_HPD` for every branch.

The **BEAST** and **MrBayes** outputs are expected to be parsed without inferences (dropped as comments) by software that supports NEXUS. **FigTree** supports parsing **BEAST**, and **MrBayes** outputs with inferences that can be used to display or annotate on the tree. But from there, extracting these data for further analysis is still challenging.

HyPhy (Pond et al., 2005) could do a number of phylogenetic analyses, including ancestral sequence reconstruction. For ancestral sequence reconstruction, these sequences and the Newick tree text are stored in NEXUS format as the major analysis output. It did not completely follow the NEXUS definition and only put the ancestral node labels in `TAXA` instead of the external node label. The `MATRIX` block contains sequence alignment of ancestral nodes which cannot be referred back to the tree stored in the `TREES` block since it does not contain node labels. Here is the sample output (to save space, only the first 72bp of alignment are shown):

```
#NEXUS

[
Generated by HYPHY 2.0020110620beta(MP) for MacOS(Universal Binary)
    on Tue Dec 23 13:52:34 2014

]

BEGIN TAXA;
    DIMENSIONS NTAX = 13;
    TAXLABELS
        'Node1' 'Node2' 'Node3' 'Node4' 'Node5' 'Node12' 'Node13' 'Node15'
            'Node18' 'Node20' 'Node22' 'Node24' 'Node26' ;
END;

BEGIN CHARACTERS;
    DIMENSIONS NCHAR = 2148;
    FORMAT
        DATATYPE = DNA

        GAP=-
        MISSING=?
        NOLABELS
    ;

MATRIX
 ATGGAAGACTTTGTGCGACAATGCTTCAATCCAATGATCGTCGAGCTTGCGGAAAAGGCAATGAAAGAATAT
 ATGGAAGACTTTGTGCGACAATGCTTCAATCCAATGATCGTCGAGCTTGCGGAAAAGGCAATGAAAGAATAT
 ATGGAAGACTTTGTGCGACAATGCTTCAATCCAATGATCGTCGAGCTTGCGGAAAAGGCAATGAAAGAATAT
 ATGGAAGACTTTGTGCGACAATGCTTCAATCCAATGATCGTCGAGCTTGCGGAAAAGGCAATGAAAGAATAT
 ATGGAAGACTTTGTGCGACAATGCTTCAATCCAATGATTGTCGAGCTTGCGGAAAAGGCAATGAAAGAATAT
 ATGGAAGACTTTGTGCGACAATGCTTCAATCCAATGATCGTCGAGCTTGCGGAAAAGGCAATGAAAGAATAT
 ATGGAAGACTTTGTGCGACAATGCTTCAATCCAATGATCGTCGAGCTTGCGGAAAAGGCAATGAAAGAATAT
 ATGGAAGACTTTGTGCGACAATGCTTCAATCCAATGATCGTCGAGCTTGCGGAAAAGGCAATGAAAGAATAT
 ATGGAAGACTTTGTGCGACAATGCTTCAATCCAATGATCGTCGAGCTTGCGGAAAAGGCAATGAAAGAATAT
 ATGGAAGACTTTGTGCGACAATGCTTCAATCCAATGATCGTCGAGCTTGCGGAAAAGGCAATGAAAGAATAT
 ATGGAAGACTTTGTGCGACAATGCTTCAATCCAATGATCGTCGAGCTTGCGGAAAAGGCAATGAAAGAATAT
 ATGGAAGACTTTGTGCGACAATGCTTCAATCCAATGATCGTCGAGCTTGCGGAAAAGGCAATGAAAGAATAT
 ATGGAAGACTTTGTGCGACAGTGCTTCAATCCAATGATCGTCGAGCTTGCGGAAAAGGCAATGAAAGAATAT
END;
```

```
BEGIN TREES;
    TREE tree = (K,N,(D,(L,(J,(G,((C,(E,O)),(H,(I,(B,(A,(F,M))))))))))));
END;
```

There are other applications that output rich information text that also contains phylogenetic trees with associated data. For example **r8s** (Sanderson, 2003) output three trees in its log file, namely `TREE`, `RATE`, and `PHYLO` for branches scaled by time, substitution rate, and absolute substitutions, respectively.

Phylogenetic Analysis by Maximum Likelihood (**PAML**) (Yang, 2007) is a package of programs for phylogenetic analyses of DNA or protein sequences. Two main programs, **BASEML** and **CODEML**, implement a variety of models. **BASEML** estimates tree topology, branch lengths, and substitution parameters using a number of nucleotide substitution models available, including JC69, K80, F81, F84, HKY85, T92, TN93, and GTR. **CODEML** estimates synonymous and non-synonymous substitution rates, likelihood ratio test of positive selection under codon substitution models (Goldman & Yang, 1994).

BASEML outputs *mlb* file that contains input sequence (taxa) alignment and a phylogenetic tree with branch length as well as substitution model and other parameters estimated. The supplementary result file, *rst*, contains sequence alignment (with ancestral sequence if it performs reconstruction of ancestral sequences) and Naive Empirical Bayes (NBE) probabilities that each site in the alignment evolved. **CODEML** outputs *mlc* file that contains tree structure and estimation of synonymous and non-synonymous substitution rates. **CODEML** also outputs a supplementary result file, *rst*, that is similar to **BASEML** except that site is defined as a codon instead of a nucleotide. Parsing these files can be tedious and would oftentimes need a number of post-processing steps and require expertise in programming (e.g., with Python[3] or Perl[4]).

Introducing square brackets is quite common for storing extra information, including *RAxML* to store bootstrap value, NHX format for annotation, jplace for edge label, and **BEAST** for evolutionary estimation, *etc.* But the positions to place square brackets are not consistent in different software and the contents employ different rules for storing associated data, these properties make it difficult to parse associated data. For most of the software, they will just ignore square brackets and only parse the tree structure if the file is compatible. Some of them contain invalid characters (e.g., curly braces in `tree` field of jplace format), and even the tree structure can't be parsed by standard parsers.

It is difficult to extract useful phylogeny/taxon-related information from the different analysis outputs produced by various evolutionary inference software, for displaying on the same phylogenetic tree and for further analysis. **FigTree** supports **BEAST** output, but not for most of the other software outputs that contain evolutionary inferences or associated data. For those output-rich text files (e.g., **r8s**, **PAML**,

[3]http://biopython.org/wiki/PAML
[4]http://bioperl.org/howtos/PAML_HOWTO.html

etc.), the tree structure cannot be parsed by any tree viewing software and users need expertise to manually extract the phylogenetic tree and other useful result data from the output file. However, such manual operation is slow and error-prone.

It was not easy to retrieve phylogenetic trees with evolutionary data from different analysis outputs of commonly used software in the field. Some of them (*e.g.*, **PAML** output and jplace file) without software or programming library to support parsing the file, while others (*e.g.*, **BEAST** and **MrBayes** output) can be parsed without evolutionary inferences as they are stored in square brackets that will be omitted as a comment by most of the software. Although **FigTree** support visualizing evolutionary statistics inferred by **BEAST** and **MrBayes**, extracting these data for further analysis is not supported. Different software packages implement different algorithms for different analyses (*e.g.*, **PAML** for d_N/d_S, **HyPhy** for ancestral sequences, and **BEAST** for skyline analysis). Therefore, in encountering the genomic sequence data, there is a desired need to efficiently and flexibly integrate different analysis inference results for comprehensive understanding, comparison, and further analysis. This motivated us to develop the programming library to parse the phylogenetic trees and data from various sources.

1.3 Getting Tree Data with treeio

Phylogenetic trees are commonly used to present evolutionary relationships of species. Information associated with taxon species/strains may be further analyzed in the context of the evolutionary history depicted by the phylogenetic tree. For example, the host information of the influenza virus strains in the tree could be studied to understand the host range of a virus lineage. Moreover, such meta-data (*e.g.*, isolation host, time, location, *etc.*) directly associated with taxon strains are also often subjected to further evolutionary or comparative phylogenetic models and analyses, to infer their dynamics associated with the evolutionary or transmission processes of the virus. All these meta-data or other phenotypic or experimental data are stored either as the annotation data associated with the nodes or branches and are often produced in inconsistent format by different analysis programs.

Getting trees into R is still limited. Newick and Nexus can be imported by several packages, including **ape**, **phylobase**. NeXML format can be parsed by **RNeXML**. However, analysis results from widely used software packages in this field are not well supported. SIMMAP output can be parsed by **phyext2** and **phytools**. Although **PHYLOCH** can import **BEAST** and **MrBayes** output, only internal node attributes were parsed and tip attributes were ignored[5]. Many other software outputs are mainly required programming expertise to import the tree with associated data. Linking external data, including experimental and clinical data, to phylogeny is another obstacle for evolution biologists.

To fill the gap that most of the tree formats or software outputs cannot be parsed within the same software/platform, an R package **treeio** (Wang et al., 2020) was

[5]https://github.com/ropensci/software-review/issues/179#issuecomment-369164110

Table 1.1: Parser functions defined in treeio

Parser function	Description
read.astral	parsing output of ASTRAL
read.beast	parsing output of BEAST
read.codeml	parsing output of CodeML (rst and mlc files)
read.codeml_mlc	parsing mlc file (output of CodeML)
read.fasta	parsing FASTA format sequence file
read.hyphy	parsing output of HYPHY
read.hyphy.seq	parsing ancestral sequences from HYPHY output
read.iqtree	parsing IQ-Tree Newick string, with the ability to parse SH-aLRT and UFBoot support values
read.jplace	parsing jplace file including the output of EPA and pplacer
read.jtree	parsing [jtree](#write-jtree) format
read.mega	parsing MEGA Nexus output
read.mega_tabular	parsing MEGA tabular output
read.mrbayes	parsing output of MrBayes
read.newick	parsing Newick string, with the ability to parse node label as support values
read.nexus	parsing standard NEXUS file (re-exported from ape)
read.nhx	parsing NHX file including the output of PHYLDOG and RevBayes
read.paml_rst	parsing rst file (output of BaseML or CodeML)
read.phylip	parsing phylip file (phylip alignment + Newick string)
read.phylip.seq	parsing multiple sequence alignment from phylip file
read.phylip.tree	parsing newick string from phylip file
read.phyloxml	parsing phyloXML file
read.r8s	parsing output of r8s
read.raxml	parsing output of RAxML
read.tree	parsing newick string (re-exported from ape)

developed for parsing various tree file formats and outputs from common evolutionary analysis software. The **treeio** package is developed with the R programming language (R Core Team, 2016). Not only the tree structure can be parsed, but also the associated data and evolutionary inferences, including NHX annotation, clock rate inferences (from **BEAST** or **r8s** (Sanderson, 2003) programs), synonymous and non-synonymous substitutions (from **CODEML**), and ancestral sequence construction (from **HyPhy**, **BASEML** or **CODEML**), *etc.*. Currently, **treeio** is able to read the following file formats: Newick, Nexus, New Hampshire eXtended format (NHX), jplace and Phylip as well as the data outputs from the following analysis programs: **ASTRAL**, **BEAST**, **EPA**, **HyPhy**, **MEGA**, **MrBayes**, **PAML**, **PHYLDOG**, **PPLACER**, **r8s**, **RAxML** and **RevBayes**, *etc.* This is made possible with the several parser functions developed in **treeio** (Table 1.1) (Wang et al., 2020).

The **treeio** package defines base classes and functions for phylogenetic tree input and output. It is an infrastructure that enables evolutionary evidence inferred by commonly used software packages to be used in R. For instance, d_N/d_S values or ancestral sequences inferred by **CODEML** (Yang, 2007), clade support values (posterior) inferred by **BEAST** (Bouckaert et al., 2014) and short read placement by **EPA** (Berger et al., 2011) and **PPLACER** (Matsen et al., 2010). These pieces of evolutionary evidence can be further analyzed in R and used to annotate a phylogenetic tree using **ggtree** (Yu et al., 2017). The growth of analysis tools and models introduces a challenge to integrate different varieties of data and analysis results from different sources for an integral analysis on the same phylogenetic tree background. The **treeio** package (Wang et al., 2020) provides a `merge_tree` function to allow combining tree data obtained from different sources. In addition, **treeio** also enables external data to be linked to a phylogenetic tree structure.

After parsing, storage of the tree structure with associated data is made through an S4 class, `treedata`, defined in the **tidytree** package. These parsed data are mapped to the tree branches and nodes inside `treedata` object, so that they can be efficiently used to visually annotate the tree using **ggtree** (Yu et al., 2017) and **ggtreeExtra** (Xu, Dai, et al., 2021). A programmable platform for phylogenetic data parsing, integration, and annotations as such makes us more easily to identify the evolutionary dynamics and correlation patterns (Figure 1.3) (Wang et al., 2020).

1.3.1 Overview of treeio

The **treeio** package (Wang et al., 2020) defined `S4` classes for storing phylogenetic trees with diverse types of associated data or covariates from different sources including analysis outputs from different software packages. It also defined corresponding parser functions for parsing phylogenetic trees with annotation data and stored them as data objects in R for further manipulation or analysis (see Table 1.1). Several accessor functions were defined to facilitate accessing the tree annotation data, including `get.fields` for obtaining annotation features available in the tree object, `get.placements` for obtaining the phylogenetic placement results (*i.e.*, the output of **PPLACER**, **EPA**, *etc.*), `get.subs` for obtaining the genetic substitutions from parent node to child node, and `get.tipseq` for getting the tip sequences.

The `S3` class, `phylo`, which was defined in **ape** (Paradis et al., 2004) package, is widely used in `R` community and many packages. As **treeio** uses `S4` class, to enable those available R packages to analyze the tree imported by **treeio**, **treeio** provides `as.phylo` function to convert **treeio**-generated tree object to `phylo` object that only contains tree structure without annotation data. In the other way, **treeio** also provides `as.treedata` function to convert `phylo` object with evolutionary analysis result (*e.g.*, bootstrap values calculated by **ape** or ancestral states inferred by **phangorn** (Schliep, 2011) *etc.*) to be stored as a `treedata` `S4` object, making it easy to map the data to the tree structure and to be visualized using **ggtree** (Yu et al., 2017).

To allow integration of different kinds of data in a phylogenetic tree, **treeio** (Wang et al., 2020) provides `merge_tree` function (details in section 2.2.1) for combining evolutionary statistics/evidence imported from different sources including those common tree files and outputs from analysis programs (Table 1.1). There is other information, such as sampling location, taxonomy information, experimental result, and evolutionary traits, *etc.* that are stored in separate files with the user-defined format. In **treeio**, we could read in these data from the users' files using standard R *IO* functions, and attach them to the tree object by the full_join methods defined in **tidytree** and **treeio** packages (see also the `%<+%` operator defined in **ggtree**). After attaching, the data will become the attributes associated with nodes or branches,

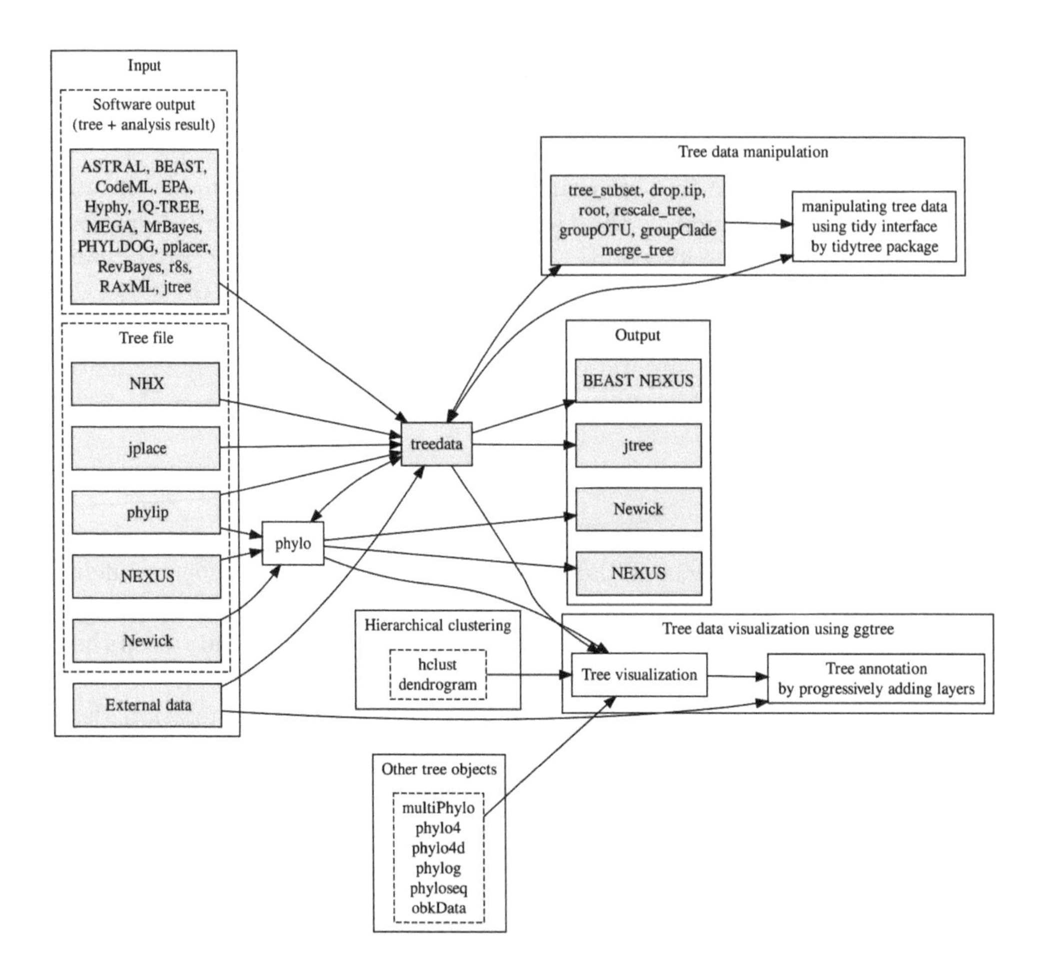

Figure 1.3: **Overview of the *treeio* package and its relations with *tidytree* and *ggtree*.** *Treeio* supports parsing a tree with data from a number of file formats and software outputs. A *treedata* object stores a phylogenetic tree with node/branch-associated data. *Treeio* provides several functions to manipulate a tree with data. Users can convert the *treedata* object into a tidy data frame (each row represents a node in the tree and each column represents a variable) and process the tree with data using the tidy interface implemented in *tidytree*. The tree can be extracted from the *treedata* object and exported to a Newick and NEXUS file or can be exported with associated data into a single file (either in the BEAST NEXUS or jtree format). Associated data stored in the *treedata* object can be used to annotate the tree using *ggtree*. In addition, *ggtree* supports a number of tree objects, including *phyloseq* for microbiome data and *obkData* for outbreak data. The *phylo*, multiPhylo (*ape* package), *phylo4*, *phylo4d* (*phylobase* package), *phylog* (*ade4* package), *phyloseq* (*phyloseq* package), and *obkData* (*OutbreakTools* package) are tree objects defined by the R community to store tree with or without domain-specific data. All these tree objects as well as hierarchical clustering results (*e.g.*, *hclust* and *dendrogram* objects) are supported by *ggtree*.

which can be compared with other data incorporated or can be visually displayed on the tree (Yu et al., 2018).

To facilitate storing complex data associated with the phylogenetic tree, **treeio** implemented `write.beast` and `write.jtree` functions to export a `treedata` object into a single file (see Chapter 3).

1.3.2 Function demonstration

1.3.2.1 Parsing BEAST output

```
file <- system.file("extdata/BEAST", "beast_mcc.tree", package="treeio")
beast <- read.beast(file)
beast

## 'treedata' S4 object that stored information
## of
##  '/home/ygc/R/library/treeio/extdata/BEAST/beast_mcc.tree'.
##
## ...@ phylo:
##
## Phylogenetic tree with 15 tips and 14 internal nodes.
##
## Tip labels:
##   A_1995, B_1996, C_1995, D_1987, E_1996, F_1997, ...
##
## Rooted; includes branch lengths.
##
## with the following features available:
##   'height', 'height_0.95_HPD', 'height_median',
## 'height_range', 'length', 'length_0.95_HPD',
## 'length_median', 'length_range', 'posterior', 'rate',
## 'rate_0.95_HPD', 'rate_median', 'rate_range'.
```

Since *%* is not a valid character in *names*, all the feature names that contain *x%* will convert to *0.x*. For example, *length_95%_HPD* will be changed to *length_0.95_HPD*.

Not only tree structure but also all the features inferred by **BEAST** will be stored in the `S4` object. These features can be used for tree annotation (Figure 5.8).

1.3.2.2 Parsing MEGA output

Molecular Evolutionary Genetics Analysis (**MEGA**) software (Kumar et al., 2016) supports exporting trees in three distinct formats: Newick, tabular, and Nexus. The Newick file can be parsed using the `read.tree` or `read.newick` functions. MEGA Nexus file is similar to BEAST Nexus and **treeio** (Wang et al., 2020) provides `read.mega` function to parse the tree.

```
file <- system.file("extdata/MEGA7", "mtCDNA_timetree.nex",
                    package = "treeio")
read.mega(file)

## 'treedata' S4 object that stored information
## of
##  '/home/ygc/R/library/treeio/extdata/MEGA7/mtCDNA_timetree.nex'.
##
## ...@ phylo:
##
## Phylogenetic tree with 7 tips and 6 internal nodes.
##
## Tip labels:
##   homo_sapiens, chimpanzee, bonobo, gorilla,
## orangutan, sumatran, ...
##
## Rooted; includes branch lengths.
##
## with the following features available:
##   'branch_length', 'data_coverage', 'rate',
## 'reltime', 'reltime_0.95_CI', 'reltime_stderr'.
```

The tabular output contains tree and associated information (divergence time in this example) in a tabular flat text file. The `read.mega_tabular` function can parse the tree with data simultaneously.

```
file <- system.file("extdata/MEGA7", "mtCDNA_timetree_tabular.txt",
                    package = "treeio")
read.mega_tabular(file)

## 'treedata' S4 object that stored information
## of
##  '/home/ygc/R/library/treeio/extdata/MEGA7/mtCDNA_timetree_tabular.
##  txt'.
##
## ...@ phylo:
##
## Phylogenetic tree with 7 tips and 6 internal nodes.
##
## Tip labels:
##   chimpanzee, bonobo, homo sapiens, gorilla,
## orangutan, sumatran, ...
## Node labels:
##   , , demoLabel2, , ,
##
## Rooted; no branch lengths.
```

```
##
## with the following features available:
##   'RelTime', 'CI_Lower', 'CI_Upper', 'Rate', 'Data
## Coverage'.
```

1.3.2.3 Parsing MrBayes output

Although the Nexus file generated by **MrBayes** is different from the output of **BEAST**, they are similar. The **treeio** package provides the `read.mrbayes()` which internally calls `read.beast()` to parse **MrBayes** outputs.

```
file <- system.file("extdata/MrBayes", "Gq_nxs.tre", package="treeio")
read.mrbayes(file)
```

```
## 'treedata' S4 object that stored information
## of
##  '/home/ygc/R/library/treeio/extdata/MrBayes/Gq_nxs.tre'.
##
## ...@ phylo:
##
## Phylogenetic tree with 12 tips and 10 internal nodes.
##
## Tip labels:
##   B_h, B_s, G_d, G_k, G_q, G_s, ...
##
## Unrooted; includes branch lengths.
##
## with the following features available:
##   'length_0.95HPD', 'length_mean', 'length_median',
## 'prob', 'prob_range', 'prob_stddev', 'prob_percent',
## 'prob+-sd'.
```

1.3.2.4 Parsing PAML output

Phylogenetic Analysis by Maximum Likelihood (**PAML**) is a package of tools for phylogenetic analyses of DNA and protein sequences using maximum likelihood. Tree search algorithms are implemented in **BASEML** and **CODEML**. The `read.paml_rst()` function provided in **treeio** can parse *rst* file from **BASEML** and **CODEML**. The only difference is the space in the sequences. For **BASEML**, every ten bases are separated by one space, while for **CODEML**, every three bases (triplet) are separated by one space.

```
brstfile <- system.file("extdata/PAML_Baseml", "rst", package="treeio")
brst <- read.paml_rst(brstfile)
brst
```

```
## 'treedata' S4 object that stored information
## of
```

```
##  '/home/ygc/R/library/treeio/extdata/PAML_Baseml/rst'.
##
## ...@ phylo:
##
## Phylogenetic tree with 15 tips and 13 internal nodes.
##
## Tip labels:
##   A, B, C, D, E, F, ...
## Node labels:
##   16, 17, 18, 19, 20, 21, ...
##
## Unrooted; includes branch lengths.
##
## with the following features available:
##   'subs', 'AA_subs'.
```

Similarly, we can parse the *rst* file from **CODEML**.

```
crstfile <- system.file("extdata/PAML_Codeml", "rst", package="treeio")
## type can be one of "Marginal" or "Joint"
crst <- read.paml_rst(crstfile, type = "Joint")
crst
```

```
## 'treedata' S4 object that stored information
## of
##  '/home/ygc/R/library/treeio/extdata/PAML_Codeml/rst'.
##
## ...@ phylo:
##
## Phylogenetic tree with 15 tips and 13 internal nodes.
##
## Tip labels:
##   A, B, C, D, E, F, ...
## Node labels:
##   16, 17, 18, 19, 20, 21, ...
##
## Unrooted; includes branch lengths.
##
## with the following features available:
##   'subs', 'AA_subs'.
```

Ancestral sequences inferred by **BASEML** or **CODEML** via marginal or joint ML reconstruction methods will be stored in the S4 object and mapped to tree nodes. **treeio** (Wang et al., 2020) will automatically determine the substitutions between the sequences at both ends of each branch. The amino acid substitution will also be determined by translating nucleotide sequences to amino acid sequences. These computed substitutions will also be stored in the S4 object for efficient tree annotation later (Figure 5.10).

CODEML infers selection pressure and estimated d_N/d_S, d_N and d_S. These pieces of information are stored in output file *mlc*, which can be parsed by the `read.codeml_mlc()` function.

```
mlcfile <- system.file("extdata/PAML_Codeml", "mlc", package="treeio")
mlc <- read.codeml_mlc(mlcfile)
mlc

## 'treedata' S4 object that stored information
## of
##  '/home/ygc/R/library/treeio/extdata/PAML_Codeml/mlc'.
##
## ...@ phylo:
##
## Phylogenetic tree with 15 tips and 13 internal nodes.
##
## Tip labels:
##   A, B, C, D, E, F, ...
## Node labels:
##   16, 17, 18, 19, 20, 21, ...
##
## Unrooted; includes branch lengths.
##
## with the following features available:
##   't', 'N', 'S', 'dN_vs_dS', 'dN', 'dS', 'N_x_dN',
## 'S_x_dS'.
```

The *rst* and *mlc* files can be parsed separately as demonstrated previously, they can also be parsed together using the `read.codeml()` function.

```
## tree can be one of "rst" or "mlc" to specify
## using tree from which file as base tree in the object
ml <- read.codeml(crstfile, mlcfile, tree = "mlc")
ml

## 'treedata' S4 object that stored information
## of
##  '/home/ygc/R/library/treeio/extdata/PAML_Codeml/rst',
##  '/home/ygc/R/library/treeio/extdata/PAML_Codeml/mlc'.
##
## ...@ phylo:
##
## Phylogenetic tree with 15 tips and 13 internal nodes.
##
## Tip labels:
##   A, B, C, D, E, F, ...
```

```
## Node labels:
##   16, 17, 18, 19, 20, 21, ...
##
## Unrooted; includes branch lengths.
##
## with the following features available:
##   'subs', 'AA_subs', 't', 'N', 'S', 'dN_vs_dS', 'dN',
## 'dS', 'N_x_dN', 'S_x_dS'.
```

All the features in both *rst* and *mlc* files are imported into a single S4 object and hence are available for further annotation and visualization. For example, we can annotate and display both d_N/d_S (from *mlc* file) and amino acid substitutions (derived from *rst* file) on the same phylogenetic tree (Yu et al., 2017).

1.3.2.5 Parsing HyPhy output

Hypothesis testing using Phylogenies (**HyPhy**) is a software package for analyzing genetic sequences. Ancestral sequences inferred by **HyPhy** are stored in the Nexus output file, which contains the tree topology and ancestral sequences. To parse this data file, users can use the `read.hyphy.seq()` function.

```
ancseq <- system.file("extdata/HYPHY", "ancseq.nex", package="treeio")
read.hyphy.seq(ancseq)
```

```
## 13 DNA sequences in binary format stored in a list.
##
## All sequences of same length: 2148
##
## Labels:
## Node1
## Node2
## Node3
## Node4
## Node5
## Node12
## ...
##
## Base composition:
##     a     c     g     t
## 0.335 0.208 0.237 0.220
## (Total: 27.92 kb)
```

To map the sequences on the tree, users should also provide an internal-node-labeled tree. If users want to determine substitution, they need to also provide tip sequences. In this case, substitutions will be determined automatically, just as we parse the output of **CODEML**.

```
nwk <- system.file("extdata/HYPHY", "labelledtree.tree", package="treeio")
tipfas <- system.file("extdata", "pa.fas", package="treeio")
hy <- read.hyphy(nwk, ancseq, tipfas)
hy
```

```
## 'treedata' S4 object that stored information
## of
##  '/home/ygc/R/library/treeio/extdata/HYPHY/labelledtree.tree'.
##
## ...@ phylo:
##
## Phylogenetic tree with 15 tips and 13 internal nodes.
##
## Tip labels:
##   K, N, D, L, J, G, ...
## Node labels:
##   Node1, Node2, Node3, Node4, Node5, Node12, ...
##
## Unrooted; includes branch lengths.
##
## with the following features available:
##   'subs', 'AA_subs'.
```

1.3.2.6 Parsing r8s output

The **r8s** package uses parametric, semi-parametric, and non-parametric methods to relax the molecular clock to allow better estimations of divergence times and evolution rates (Sanderson, 2003). It outputs three trees in a log file, namely, *TREE*, *RATO*, and *PHYLO* for time tree, rate tree, and absolute substitution tree, respectively.

The time tree is scaled by divergence time, the rate tree is scaled by substitution rate and the absolute substitution tree is scaled by the absolute number of substitutions. After parsing the file, all these three trees are stored in a *multiPhylo* object (Figure 4.15).

```
r8s <- read.r8s(system.file("extdata/r8s", "H3_r8s_output.log",
                package="treeio"))
r8s
```

```
## 3 phylogenetic trees
```

1.3.2.7 Parsing output of RAxML bootstrap analysis

RAxML bootstrapping analysis outputs a Newick tree text that is not standard, as it stores bootstrap values inside square brackets after branch lengths. This file usually cannot be parsed by a traditional Newick parser, such as `ape::read.tree()`. The function `read.raxml()` can read such files and store the bootstrap as an additional feature, which can be used to display on the tree or used to color tree branches, *etc.*

```
raxml_file <- system.file("extdata/RAxML",
                          "RAxML_bipartitionsBranchLabels.H3",
                          package="treeio")
raxml <- read.raxml(raxml_file)
raxml

## 'treedata' S4 object that stored information
## of
##  '/home/ygc/R/library/treeio/extdata/RAxML/RAxML_
##      bipartitionsBranchLabels.H3'.
##
## ...@ phylo:
##
## Phylogenetic tree with 64 tips and 62 internal nodes.
##
## Tip labels:
##   A_Hokkaido_M1_2014_H3N2_2014,
## A_Czech_Republic_1_2014_H3N2_2014,
## FJ532080_A_California_09_2008_H3N2_2008,
## EU199359_A_Pennsylvania_05_2007_H3N2_2007,
## EU857080_A_Hong_Kong_CUHK69904_2006_H3N2_2006,
## EU857082_A_Hong_Kong_CUHK7047_2005_H3N2_2005, ...
##
## Unrooted; includes branch lengths.
##
## with the following features available:
##   'bootstrap'.
```

1.3.2.8 Parsing NHX tree

NHX (New Hampshire eXtended) format is an extension of Newick by introducing NHX tags. NHX is commonly used in phylogenetics software, including **PHYLDOG** (Boussau et al., 2013), **RevBayes** (Höhna et al., 2014), for storing statistical inferences. The following codes imported an NHX tree with associated data inferred by **PHYLDOG** (Figure 3.1A).

```
nhxfile <- system.file("extdata/NHX", "phyldog.nhx", package="treeio")
nhx <- read.nhx(nhxfile)
nhx

## 'treedata' S4 object that stored information
## of
##  '/home/ygc/R/library/treeio/extdata/NHX/phyldog.nhx'.
##
```

```
## ...@ phylo:
##
## Phylogenetic tree with 16 tips and 15 internal nodes.
##
## Tip labels:
##   Prayidae_D27SS7@2825365, Kephyes_ovata@2606431,
## Chuniphyes_multidentata@1277217,
## Apolemia_sp_@1353964, Bargmannia_amoena@263997,
## Bargmannia_elongata@946788, ...
##
## Rooted; includes branch lengths.
##
## with the following features available:
##   'Ev', 'ND', 'S'.
```

1.3.2.9 Parsing Phylip tree

Phylip format contains multiple sequence alignment of taxa in Phylip sequence format with corresponding Newick tree text that was built from taxon sequences. Multiple sequence alignment can be sorted based on the tree structure and displayed at the right-hand side of the tree using **ggtree** through the `msaplot()` function or in combining with the **ggmsa** package (see also Basic Protocol 5 of (Yu, 2020)).

```
phyfile <- system.file("extdata", "sample.phy", package="treeio")
phylip <- read.phylip(phyfile)
phylip
```

```
## 'treedata' S4 object that stored information
## of
##  '/home/ygc/R/library/treeio/extdata/sample.phy'.
##
## ...@ phylo:
##
## Phylogenetic tree with 15 tips and 13 internal nodes.
##
## Tip labels:
##   K, N, D, L, J, G, ...
##
## Unrooted; no branch lengths.
```

1.3.2.10 Parsing EPA and pplacer output

EPA (Berger et al., 2011) and **PPLACER** (Matsen et al., 2010) have a common output file format, `jplace`, which can be parsed by the `read.jplace()` function.

```
jpf <- system.file("extdata/EPA.jplace",  package="treeio")
jp <- read.jplace(jpf)
```

```
print(jp)

## 'treedata' S4 object that stored information
## of
##  '/home/ygc/R/library/treeio/extdata/EPA.jplace'.
##
## ...@ phylo:
##
## Phylogenetic tree with 493 tips and 492 internal nodes.
##
## Tip labels:
##   CIR000447A, CIR000479, CIR000078, CIR000083,
## CIR000070, CIR000060, ...
##
## Rooted; includes branch lengths.
##
## with the following features available:
##   'nplace'.
```

The number of evolutionary placement on each branch will be calculated and stored as the *nplace* feature, which can be mapped to line size and/or color using **ggtree** (Yu et al., 2017).

1.3.2.11 Parsing jtree format

The *jtree* is a JSON-based format that was defined in this **treeio** package (Wang et al., 2020) to support tree data interchange (see session 3.3). Phylogenetic tree with associated data can be exported to a single *jtree* file using the `write.jtree()` function. The *jtree* can be easily parsed using any JSON parser. The *jtree* format contains three keys: tree, data, and meta-data. The tree value contains tree text extended from Newick tree format by putting the edge number in curly braces after branch length. The data value contains node/branch-specific data, while the meta-data value contains additional meta information.

```
jtree_file <- tempfile(fileext = '.jtree')
write.jtree(beast, file = jtree_file)
read.jtree(file = jtree_file)

## 'treedata' S4 object that stored information
## of
##  '/tmp/RtmpiTfY9K/file115c89ababf4.jtree'.
##
## ...@ phylo:
##
## Phylogenetic tree with 15 tips and 14 internal nodes.
##
```

Table 1.2: Conversion of tree-like object to phylo or treedata object

Convert function	Supported object	Description
as.phylo	ggtree	convert ggtree object to phylo object
	igraph	convert igraph object (only tree graph supported) to phylo object
	phylo4	convert phylo4 object to phylo object
	pvclust	convert pvclust object to phylo object
	treedata	convert treedata object to phylo object
as.treedata	ggtree	convert ggtree object to treedata object
	phylo4	convert phylo4 object to treedata object
	phylo4d	convert phylo4d object to treedata object
	pml	convert pml object to treedata object
	pvclust	convert pvclust object to treedata object

```
## Tip labels:
##   K_2013, N_2010, D_1987, L_1980, J_1983, G_1992, ...
##
## Rooted; includes branch lengths.
##
## with the following features available:
##   'height', 'height_0.95_HPD', 'height_range',
## 'length', 'length_0.95_HPD', 'length_median',
## 'length_range', 'rate', 'rate_0.95_HPD',
## 'rate_median', 'rate_range', 'height_median',
## 'posterior'.
```

1.3.3 Converting other tree-like objects to `phylo` or `treedata` objects

To extend the application scopes of **treeio**, **tidytree** and **ggtree**, **treeio** (Wang et al., 2020) provides several `as.phylo` and `as.treedata` methods to convert other tree-like objects, such as `phylo4d` and `pml`, to `phylo` or `treedata` object. So that users can easily map associated data to the tree structure, export a tree with/without data to a single file, manipulate and visualize a tree with/without data. These convert functions (Table 1.2) create the possibility of using **tidytree** to process tree using tidy interface and **ggtree** to visualize tree using the grammar of graphic syntax.

Here, we used `pml` object which was defined in the **phangorn** package, as an example. The `pml()` function computes the likelihood of a phylogenetic tree given a sequence alignment and a model and the `optim.pml()` function optimizes different model parameters. The output is a `pml` object, and it can be converted to a `treedata` object using `as.treedata` provided by **treeio** (Wang et al., 2020). The amino acid substitution (ancestral sequence estimated by `pml`) that stored in the `treedata` object can be visualized using **ggtree** as demonstrated in Figure 1.4.

```
library(phangorn)
treefile <- system.file("extdata", "pa.nwk", package="treeio")
tre <- read.tree(treefile)
tipseqfile <- system.file("extdata", "pa.fas", package="treeio")
tipseq <- read.phyDat(tipseqfile,format="fasta")
fit <- pml(tre, tipseq, k=4)
fit <- optim.pml(fit, optNni=FALSE, optBf=T, optQ=T,
                 optInv=T, optGamma=T, optEdge=TRUE,
                 optRooted=FALSE, model = "GTR",
                 control = pml.control(trace =0))

pmltree <- as.treedata(fit)
ggtree(pmltree) + geom_text(aes(x=branch, label=AA_subs, vjust=-.5))
```

1.3.4 Getting information from treedata object

After the tree was imported, users may want to extract information stored in the `treedata` object. **treeio** provides several accessor methods to extract tree structure, features/attributes that stored in the object, and their corresponding values.

The `get.tree()` or `as.phylo()` methods can convert the `treedata` object to a `phylo` object which is the fundamental tree object in the R community and many packages work with `phylo` object.

```
beast_file <- system.file("examples/MCC_FluA_H3.tree", package="ggtree")
beast_tree <- read.beast(beast_file)
# or get.tree
as.phylo(beast_tree)
```

```
beast_file <- system.file("examples/MCC_FluA_H3.tree", package="ggtree")
beast_tree <- read.beast(beast_file)
# or get.tree
print(as.phylo(beast_tree), printlen=3)
```

```
##
## Phylogenetic tree with 76 tips and 75 internal nodes.
##
## Tip labels:
##   A/Hokkaido/30-1-a/2013, A/New_York/334/2004, A/New_York/463/2005,
##    ...
## Rooted; includes branch lengths.
```

The `get.fields` method returns a vector of features/attributes stored in the object and associated with the phylogeny.

```
get.fields(beast_tree)
```

```
##  [1] "height"          "height_0.95_HPD"
```

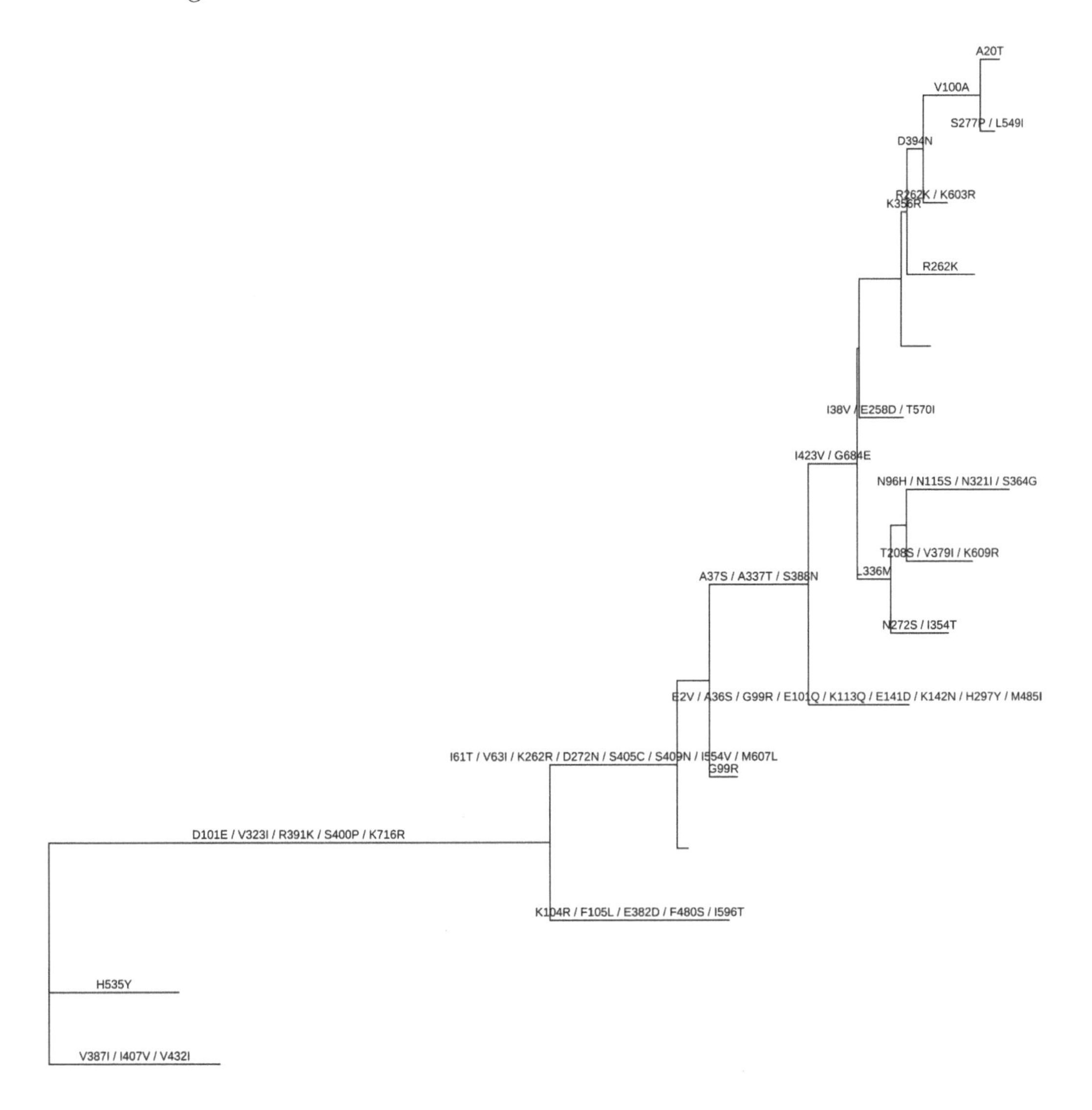

Figure 1.4: **Converting `pml` object to `treedata` object.** This allows using **tidytree** to process the tree data as well as using **ggtree** and **ggtreeExtra** to visualize the tree with associated data.

```
##  [3] "height_median"   "height_range"
##  [5] "length"          "length_0.95_HPD"
##  [7] "length_median"   "length_range"
##  [9] "posterior"       "rate"
## [11] "rate_0.95_HPD"   "rate_median"
## [13] "rate_range"
```

The `get.data` method returns a tibble of all the associated data.

```
get.data(beast_tree)
```

```
## # A tibble: 151 x 14
##    height height_0.95_HPD height_median height_range
##     <dbl> <list>                  <dbl> <list>
##  1   19   <dbl [2]>                19   <dbl [2]>
##  2   17   <dbl [2]>                17   <dbl [2]>
##  3   14   <dbl [2]>                14   <dbl [2]>
##  4   12   <dbl [2]>                12   <dbl [2]>
##  5    9   <dbl [2]>                 9   <dbl [2]>
##  6   10   <dbl [2]>                10   <dbl [2]>
##  7   10   <dbl [2]>                10   <dbl [2]>
##  8   10.8 <dbl [2]>                10.8 <dbl [2]>
##  9    9   <dbl [2]>                 9   <dbl [2]>
## 10    9   <dbl [2]>                 9   <dbl [2]>
## # ... with 141 more rows, and 10 more variables:
## #   length <dbl>, length_0.95_HPD <list>,
## #   length_median <dbl>, length_range <list>,
## #   posterior <dbl>, rate <dbl>, rate_0.95_HPD <list>,
## #   rate_median <dbl>, rate_range <list>, node <int>
```

If users are only interested in a subset of the features/attributes returned by `get.fields`, they can extract the information from the output of `get.data` or directly subset the data by `[` or `[[`.

```
beast_tree[, c("node", "height")]
```

```
## # A tibble: 151 x 2
##     node height
##    <int>  <dbl>
##  1    10   19
##  2     9   17
##  3    36   14
##  4    31   12
##  5    29    9
##  6    28   10
##  7    39   10
##  8    90   10.8
##  9    16    9
## 10     2    9
## # ... with 141 more rows
```

```
head(beast_tree[["height_median"]])
```

```
## height_median1 height_median2 height_median3
##             19             17             14
```

```
## height_median4 height_median5 height_median6
##             12              9             10
```

1.4 Summary

Software tools for inferring molecular evolution (*e.g.*, ancestral states, molecular dating and selection pressure, *etc.*) are proliferating, but there is no single data format that is used by all different programs and capable to store different types of phylogenetic data. Most of the software packages have their unique output formats and these formats are not compatible with each other. Parsing software outputs is challenging, which restricts the joint analysis using different tools. The **treeio** package (Wang et al., 2020) provides a set of functions (Table 1.1) for parsing various types of phylogenetic data files and a set of converters (Table 1.2) to convert tree-like objects to phylo or treedata objects. These phylogenetic data can be integrated that allow further exploration and comparison. To date, most software tools in the field of molecular evolution are isolated and often not fully compatible with each other's input and output files. These software tools are designed to do their analysis and the outputs are often not readable in other software. No tools have been designed to unify the inference data from different analysis programs. Efficient incorporation of data from different inference methods can enhance the comparison and understanding of the study target, which may help discover new systematic patterns and generate new hypotheses.

As phylogenetic trees are growing in their application to identify patterns in an evolutionary context, more different disciplines are employing phylogenetic trees in their research. For example, spatial ecologists may map the geographical positions of the organisms to their phylogenetic trees to understand the biogeography of the species (Schön et al., 2015); disease epidemiologists may incorporate the pathogen sampling time and locations into the phylogenetic analysis to infer the disease transmission dynamics in spatiotemporal space (Y.-Q. He et al., 2013); microbiologists may determine the pathogenicity of different pathogen strains and map them into their phylogenetic trees to identify the genetic determinants of the pathogenicity (Bosi et al., 2016); genomic scientists may use the phylogenetic trees to help taxonomically classify their metagenomic sequence data (Gupta & Sharma, 2015). A robust tool such as **treeio** to import and map different types of data into the phylogenetic tree is important to facilitate these phylogenetics-related research, or 'phylodynamics'. Such tools could also help integrate different meta-data (time, geography, genotype, epidemiological information) and analysis results (selective pressure, evolutionary rates) at the highest level and provide a comprehensive understanding of the study organisms. In the field of influenza research, there have been such attempts of studying the phylodynamics of the influenza virus by mapping different meta-data and analysis results on the same phylogenetic tree and evolutionary timescale (Lam et al., 2015).

Chapter 2

Manipulating Tree with Data

2.1 Manipulating Tree Data Using Tidy Interface

All the tree data parsed/merged by **treeio** (Wang et al., 2020) can be converted to a tidy data frame using the **tidytree** package. The **tidytree** package provides tidy interfaces to manipulate trees with associated data. For instance, external data can be linked to phylogeny or evolutionary data obtained from different sources can be merged using tidyverse verbs. After the tree data was processed, it can be converted back to a `treedata` object and exported to a single tree file, further analyzed in R or visualized using **ggtree** (Yu et al., 2017) and **ggtreeExtra** (Xu, Dai, et al., 2021).

2.1.1 The `phylo` object

The `phylo` class defined in the **ape** package (Paradis et al., 2004) is fundamental for phylogenetic analysis in R. Most of the R packages in this field rely extensively on the `phylo` object. The **tidytree** package provides `as_tibble` method to convert the `phylo` object to a tidy data frame, a `tbl_tree` object.

```
library(ape)
set.seed(2017)
tree <- rtree(4)
tree
```

```
##
## Phylogenetic tree with 4 tips and 3 internal nodes.
##
## Tip labels:
##   t4, t1, t3, t2
##
## Rooted; includes branch lengths.
```

```
x <- as_tibble(tree)
x
```

```
## # A tibble: 7 x 4
##   parent  node branch.length label
##    <int> <int>         <dbl> <chr>
## 1      5     1       0.435   t4
## 2      7     2       0.674   t1
## 3      7     3       0.00202 t3
## 4      6     4       0.0251  t2
## 5      5     5      NA       <NA>
## 6      5     6       0.472   <NA>
## 7      6     7       0.274   <NA>
```

The `tbl_tree` object can be converted back to a `phylo` object using the `as.phylo()` method.

```
as.phylo(x)
```

```
##
## Phylogenetic tree with 4 tips and 3 internal nodes.
##
## Tip labels:
##   t4, t1, t3, t2
##
## Rooted; includes branch lengths.
```

Using `tbl_tree` object makes tree and data manipulation more effective and easier (see also the example in FAQ). For example, we can link evolutionary trait to phylogeny using the **dplyr** verbs `full_join()`:

```
d <- tibble(label = paste0('t', 1:4),
            trait = rnorm(4))

y <- full_join(x, d, by = 'label')
y
```

```
## # A tibble: 7 x 5
##   parent  node branch.length label  trait
##    <int> <int>         <dbl> <chr>  <dbl>
## 1      5     1       0.435   t4     0.943
## 2      7     2       0.674   t1    -0.171
## 3      7     3       0.00202 t3     0.570
## 4      6     4       0.0251  t2    -0.283
## 5      5     5      NA       <NA>  NA
## 6      5     6       0.472   <NA>  NA
## 7      6     7       0.274   <NA>  NA
```

2.1.2 The treedata object

The **tidytree** package defines `treedata` class to store a phylogenetic tree with associated data. After mapping external data to the tree structure, the `tbl_tree` object can be converted to a `treedata` object.

```
as.treedata(y)
```

```
## 'treedata' S4 object'.
##
## ...@ phylo:
##
## Phylogenetic tree with 4 tips and 3 internal nodes.
##
## Tip labels:
##   t4, t1, t3, t2
##
## Rooted; includes branch lengths.
##
## with the following features available:
##   'trait'.
```

The `treedata` class is used in the treeio package (Wang et al., 2020) to store evolutionary evidence inferred by commonly used software (**BEAST**, **EPA**, **HyPhy**, **MrBayes**, **PAML**, **PHYLDOG**, **PPLACER**, **r8s**, **RAxML**, and **RevBayes**, etc.) (see details in Chapter 1).

The **tidytree** package also provides the `as_tibble()` method to convert a `treedata` object to a tidy data frame. The phylogenetic tree structure and the evolutionary inferences were stored in the `tbl_tree` object, making it consistent and easier for manipulating evolutionary statistics inferred by different software as well as linking external data to the same tree structure.

```
y %>% as.treedata %>% as_tibble
```

```
## # A tibble: 7 x 5
##   parent  node branch.length label  trait
##    <int> <int>         <dbl> <chr>  <dbl>
## 1      5     1       0.435   t4     0.943
## 2      7     2       0.674   t1    -0.171
## 3      7     3       0.00202 t3     0.570
## 4      6     4       0.0251  t2    -0.283
## 5      5     5      NA       <NA>  NA
## 6      5     6       0.472   <NA>  NA
## 7      6     7       0.274   <NA>  NA
```

2.1.3 Access related nodes

The **dplyr** verbs can be applied to `tbl_tree` directly to manipulate tree data. In addition, **tidytree** provides several verbs to filter related nodes, including `child()`, `parent()`, `offspring()`, `ancestor()`, `sibling()` and `MRCA()`.

These verbs accept a `tbl_tree` object and a selected node which can be node number or label.

```
child(y, 5)
```

```
## # A tibble: 2 x 5
##   parent  node branch.length label  trait
##    <int> <int>         <dbl> <chr>  <dbl>
## 1      5     1         0.435 t4     0.943
## 2      5     6         0.472 <NA>  NA
```

```
parent(y, 2)
```

```
## # A tibble: 1 x 5
##   parent  node branch.length label trait
##    <int> <int>         <dbl> <chr> <dbl>
## 1      6     7         0.274 <NA>     NA
```

```
offspring(y, 5)
```

```
## # A tibble: 6 x 5
##   parent  node branch.length label  trait
##    <int> <int>         <dbl> <chr>  <dbl>
## 1      5     1       0.435   t4     0.943
## 2      7     2       0.674   t1    -0.171
## 3      7     3       0.00202 t3     0.570
## 4      6     4       0.0251  t2    -0.283
## 5      5     6       0.472   <NA>  NA
## 6      6     7       0.274   <NA>  NA
```

```
ancestor(y, 2)
```

```
## # A tibble: 3 x 5
##   parent  node branch.length label trait
##    <int> <int>         <dbl> <chr> <dbl>
## 1      5     5        NA     <NA>     NA
## 2      5     6         0.472 <NA>     NA
## 3      6     7         0.274 <NA>     NA
```

```
sibling(y, 2)
```

```
## # A tibble: 1 x 5
##   parent  node branch.length label trait
```

```
##     <int> <int>         <dbl> <chr> <dbl>
## 1       7     3       0.00202 t3    0.570
```

```
MRCA(y, 2, 3)
```

```
## # A tibble: 1 x 5
##    parent  node branch.length label trait
##     <int> <int>         <dbl> <chr> <dbl>
## 1       6     7         0.274 <NA>     NA
```

All these methods are also implemented in **treeio** for working with `phylo` and `treedata` objects. You can try accessing related nodes using the tree object. For instance, the following command will output child nodes of the selected internal node 5:

```
child(tree, 5)
```

```
## [1] 1 6
```

Beware that the methods for tree objects output relevant node numbers, while the methods for `tbl_tree` object output a `tibble` object that contains related information.

2.2 Data Integration

2.2.1 Combining tree data

The **treeio** package (Wang et al., 2020) serves as an infrastructure that enables various types of phylogenetic data inferred from common analysis programs to be imported and used in R. For instance, d_N/d_S or ancestral sequences estimated by **CODEML**, and clade support values (posterior) inferred by **BEAST/MrBayes**. In addition, **treeio** supports linking external data to phylogeny. It brings these external phylogenetic data (either from software output or external sources) to the R community and makes it available for further analysis in R. Furthermore, **treeio** can combine multiple phylogenetic trees into one with their node/branch-specific attribute data. Essentially, as a result, one such attribute (*e.g.*, substitution rate) can be mapped to another attribute (*e.g.*, d_N/d_S) of the same node/branch for comparison and further computations (Yu et al., 2017; Yu et al., 2018).

A previously published dataset, seventy-six H3 hemagglutinin gene sequences of a lineage containing swine and human influenza A viruses (Liang et al., 2014), was used here to demonstrate the utilities of comparing evolutionary statistics inferred by different software. The dataset was re-analyzed by **BEAST** for timescale estimation and **CODEML** for synonymous and non-synonymous substitution estimation. In this example, we first parsed the outputs from **BEAST** using the `read.beast()` function and from **CODEML** using the `read.codeml()` function into two `treedata` objects. Then these two objects containing separate sets of node/branch-specific data were merged via the `merge_tree()` function.

```
beast_file <- system.file("examples/MCC_FluA_H3.tree", package="ggtree")
rst_file <- system.file("examples/rst", package="ggtree")
mlc_file <- system.file("examples/mlc", package="ggtree")
beast_tree <- read.beast(beast_file)
codeml_tree <- read.codeml(rst_file, mlc_file)

merged_tree <- merge_tree(beast_tree, codeml_tree)
merged_tree
```

```
## 'treedata' S4 object that stored information
## of
##  '/home/ygc/R/library/ggtree/examples/MCC_FluA_H3.tree',
##  '/home/ygc/R/library/ggtree/examples/rst',
##  '/home/ygc/R/library/ggtree/examples/mlc'.
##
## ...@ phylo:
##
## Phylogenetic tree with 76 tips and 75 internal nodes.
##
## Tip labels:
##   A/Hokkaido/30-1-a/2013, A/New_York/334/2004,
## A/New_York/463/2005, A/New_York/452/1999,
## A/New_York/238/2005, A/New_York/523/1998, ...
##
## Rooted; includes branch lengths.
##
## with the following features available:
##   'height', 'height_0.95_HPD', 'height_median',
## 'height_range', 'length', 'length_0.95_HPD',
## 'length_median', 'length_range', 'posterior', 'rate',
## 'rate_0.95_HPD', 'rate_median', 'rate_range', 'subs',
## 'AA_subs', 't', 'N', 'S', 'dN_vs_dS', 'dN', 'dS',
## 'N_x_dN', 'S_x_dS'.
```

After merging the `beast_tree` and `codeml_tree` objects, all node/branch-specific data imported from **BEAST** and **CODEML** output files are all available in the `merged_tree` object. The tree object was converted to a tidy data frame using the tidytree package and visualized as hexbin scatterplots of d_N/d_S, d_N, and d_S inferred by **CODEML** vs. *rate* (substitution rate in a unit of substitutions/site/year) inferred by **BEAST** on the same branches.

```
library(dplyr)
df <- merged_tree %>%
  as_tibble() %>%
  select(dN_vs_dS, dN, dS, rate) %>%
  subset(dN_vs_dS >=0 & dN_vs_dS <= 1.5) %>%
```

```
  tidyr::gather(type, value, dN_vs_dS:dS)
df$type[df$type == 'dN_vs_dS'] <- 'dN/dS'
df$type <- factor(df$type, levels=c("dN/dS", "dN", "dS"))
ggplot(df, aes(rate, value)) + geom_hex() +
  facet_wrap(~type, scale='free_y')
```

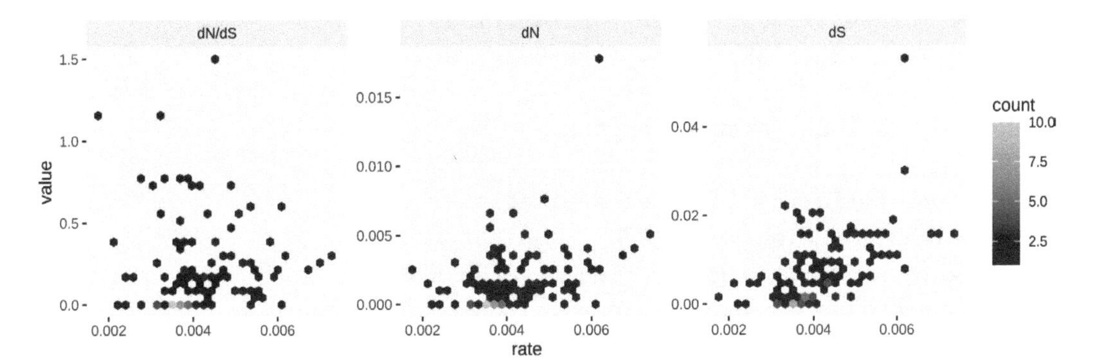

Figure 2.1: **Correlation of d_N/d_S, d_N, and d_S vs. substitution rate.** After merging the *BEAST* and *CodeML* outputs, the branch-specific estimates (substitution rate, d_N/d_S , d_N, and d_S) from the two analysis programs are compared on the same branch basis. The associations of d_N/d_S, d_N, and d_S vs. *rate* are visualized in hexbin scatter plots.

The output is illustrated in Figure 2.1. We can then test the association of these node/branch-specific data using Pearson correlation, which in this case showed that d_N and d_S, but not d_N/d_S are significantly (p-values) associated with *rate*.

Using the `merge_tree()` function, we are able to compare analysis results using an identical model from different software packages or different models using different or identical software. It also allows users to integrate different analysis findings from different software packages. Merging tree data is not restricted to software findings, associating external data to analysis findings is also granted. The `merge_tree()` function is chainable and allows several tree objects to be merged into one.

```
phylo <- as.phylo(beast_tree)
N <- Nnode2(phylo)
d <- tibble(node = 1:N, fake_trait = rnorm(N), another_trait = runif(N))
fake_tree <- treedata(phylo = phylo, data = d)
triple_tree <- merge_tree(merged_tree, fake_tree)
triple_tree

## 'treedata' S4 object that stored information
## of
##  '/home/ygc/R/library/ggtree/examples/MCC_FluA_H3.tree',
##  '/home/ygc/R/library/ggtree/examples/rst',
##  '/home/ygc/R/library/ggtree/examples/mlc'.
##
```

```
## ...@ phylo:
##
## Phylogenetic tree with 76 tips and 75 internal nodes.
##
## Tip labels:
##   A/Hokkaido/30-1-a/2013, A/New_York/334/2004,
## A/New_York/463/2005, A/New_York/452/1999,
## A/New_York/238/2005, A/New_York/523/1998, ...
##
## Rooted; includes branch lengths.
##
## with the following features available:
##   'height', 'height_0.95_HPD', 'height_median',
## 'height_range', 'length', 'length_0.95_HPD',
## 'length_median', 'length_range', 'posterior', 'rate',
## 'rate_0.95_HPD', 'rate_median', 'rate_range', 'subs',
## 'AA_subs', 't', 'N', 'S', 'dN_vs_dS', 'dN', 'dS',
## 'N_x_dN', 'S_x_dS', 'fake_trait', 'another_trait'.
##
## # The associated data tibble abstraction: 151 x 28
## # The 'node', 'label' and 'isTip' are from the phylo tree.
##     node label              isTip height height_0.95_HPD
##    <int> <chr>              <lgl>  <dbl> <list>
##  1     1 A/Hokkaido/30-1-~ TRUE       0 <lgl [1]>
##  2     2 A/New_York/334/2~ TRUE       9 <dbl [2]>
##  3     3 A/New_York/463/2~ TRUE       8 <dbl [2]>
##  4     4 A/New_York/452/1~ TRUE      14 <dbl [2]>
##  5     5 A/New_York/238/2~ TRUE       8 <dbl [2]>
##  6     6 A/New_York/523/1~ TRUE      15 <dbl [2]>
##  7     7 A/New_York/435/2~ TRUE      13 <dbl [2]>
##  8     8 A/New_York/582/1~ TRUE      17 <dbl [2]>
##  9     9 A/New_York/603/1~ TRUE      17 <dbl [2]>
## 10    10 A/New_York/657/1~ TRUE      19 <dbl [2]>
## # ... with 141 more rows, and 23 more variables:
## #   height_median <dbl>, height_range <list>,
## #   length <dbl>, length_0.95_HPD <list>,
## #   length_median <dbl>, length_range <list>,
## #   posterior <dbl>, rate <dbl>, rate_0.95_HPD <list>,
## #   rate_median <dbl>, rate_range <list>, subs <chr>,
## #   AA_subs <chr>, t <dbl>, N <dbl>, S <dbl>, ...
```

The `triple_tree` object shown above contains analysis results obtained from **BEAST** and **CODEML**, and evolutionary traits from external sources. All these pieces of information can be used to annotate the tree using **ggtree** (Yu et al., 2017) and **ggtreeExtra** (Xu, Dai, et al., 2021).

2.2.2 Linking external data to phylogeny

In addition to analysis findings that are associated with the tree as demonstrated above, there is a wide range of heterogeneous data, including phenotypic data, experimental data, and clinical data, *etc.*, that need to be integrated and linked to phylogeny. For example, in the study of viral evolution, tree nodes may be associated with epidemiological information, such as location, age, and subtype. Functional annotations may need to be mapped onto gene trees for comparative genomics studies. To facilitate data integration, **treeio** provides `full_join()` methods to link external data to phylogeny and store it in either `phylo` or `treedata` object. Beware that linking external data to a `phylo` object will produce a `treedata` object to store the input `phylo` with associated data. The `full_join` methods can also be used at tidy data frame level (*i.e.*, `tbl_tree` object described previously) and at `ggtree` level (described in Chapter 7) (Yu et al., 2018).

The following example calculated bootstrap values and merged those values with the tree (a `phylo` object) by matching their node numbers.

```
library(ape)
data(woodmouse)
d <- dist.dna(woodmouse)
tr <- nj(d)
bp <- boot.phylo(tr, woodmouse, function(x) nj(dist.dna(x)))

## Running bootstraps:       100 / 100
## Calculating bootstrap values... done.
```

```
bp2 <- tibble(node=1:Nnode(tr) + Ntip(tr), bootstrap = bp)
full_join(tr, bp2, by="node")

## 'treedata' S4 object'.
##
## ...@ phylo:
##
## Phylogenetic tree with 15 tips and 13 internal nodes.
##
## Tip labels:
##   No305, No304, No306, No0906S, No0908S, No0909S, ...
##
## Unrooted; includes branch lengths.
##
## with the following features available:
##   'bootstrap'.
##
## # The associated data tibble abstraction: 28 x 4
## # The 'node', 'label' and 'isTip' are from the phylo tree.
##     node label   isTip bootstrap
```

```
##     <int> <chr>   <lgl>     <int>
##  1     1 No305   TRUE         NA
##  2     2 No304   TRUE         NA
##  3     3 No306   TRUE         NA
##  4     4 No0906S TRUE         NA
##  5     5 No0908S TRUE         NA
##  6     6 No0909S TRUE         NA
##  7     7 No0910S TRUE         NA
##  8     8 No0912S TRUE         NA
##  9     9 No0913S TRUE         NA
## 10    10 No1103S TRUE         NA
## # ... with 18 more rows
```

Another example demonstrates merging evolutionary traits with the tree (a `treedata` object) by matching their tip labels.

```
file <- system.file("extdata/BEAST", "beast_mcc.tree", package="treeio")
beast <- read.beast(file)
x <- tibble(label = as.phylo(beast)$tip.label, trait = rnorm(Ntip(beast)))
full_join(beast, x, by="label")
```

```
## 'treedata' S4 object that stored information
## of
##  '/home/ygc/R/library/treeio/extdata/BEAST/beast_mcc.tree'.
##
## ...@ phylo:
##
## Phylogenetic tree with 15 tips and 14 internal nodes.
##
## Tip labels:
##   A_1995, B_1996, C_1995, D_1987, E_1996, F_1997, ...
##
## Rooted; includes branch lengths.
##
## with the following features available:
##   'height', 'height_0.95_HPD', 'height_median',
## 'height_range', 'length', 'length_0.95_HPD',
## 'length_median', 'length_range', 'posterior', 'rate',
## 'rate_0.95_HPD', 'rate_median', 'rate_range', 'trait'.
##
## # The associated data tibble abstraction: 29 x 17
## # The 'node', 'label' and 'isTip' are from the phylo tree.
##     node label  isTip height height_0.95_HPD
##    <int> <chr>  <lgl>  <dbl> <list>
##  1     1 A_1995 TRUE      18 <dbl [2]>
##  2     2 B_1996 TRUE      17 <dbl [2]>
```

```
## 3        3 C_1995 TRUE       18 <dbl [2]>
## 4        4 D_1987 TRUE       26 <dbl [2]>
## 5        5 E_1996 TRUE       17 <dbl [2]>
## 6        6 F_1997 TRUE       16 <dbl [2]>
## 7        7 G_1992 TRUE       21 <dbl [2]>
## 8        8 H_1992 TRUE       21 <dbl [2]>
## 9        9 I_1994 TRUE       19 <dbl [2]>
## 10      10 J_1983 TRUE       30 <dbl [2]>
## # ... with 19 more rows, and 12 more variables:
## #   height_median <dbl>, height_range <list>,
## #   length <dbl>, length_0.95_HPD <list>,
## #   length_median <dbl>, length_range <list>,
## #   posterior <dbl>, rate <dbl>, rate_0.95_HPD <list>,
## #   rate_median <dbl>, rate_range <list>, trait <dbl>
```

Manipulating tree objects is frustrated with the fragmented functions available for working with `phylo` objects, not to mention linking external data to the phylogeny structure. With the **treeio** package (Wang et al., 2020), it is easy to combine tree data from various sources. In addition, with the **tidytree** package, manipulating trees is easier using the tidy data principles and consistent with tools already in wide use, including **dplyr**, **tidyr**, **ggplot2**, and **ggtree** (Yu et al., 2017).

2.2.3 Grouping taxa

The `groupOTU()` and `groupClade()` methods are designed for adding taxa grouping information to the input tree object. The methods were implemented in **tidytree**, **treeio**, and **ggtree** respectively to support adding grouping information for the `tbl_tree`, `phylo` and `treedata`, and `ggtree` objects. This grouping information can be used directly in tree visualization (*e.g.*, coloring a tree based on grouping information) with **ggtree** (Figure 6.5).

2.2.3.1 groupClade

The `groupClade()` method accepts an internal node or a vector of internal nodes to add grouping information of selected clade/clades.

```
nwk <- '(((((((A:4,B:4):6,C:5):8,D:6):3,E:21):10,((F:4,G:12):14,H:8):13):
        13,((I:5,J:2):30,(K:11,L:11):2):17):4,M:56);'
tree <- read.tree(text=nwk)

groupClade(as_tibble(tree), c(17, 21))
```

```
## # A tibble: 25 x 5
##    parent  node branch.length label group
##     <int> <int>         <dbl> <chr> <fct>
## 1      20     1             4 A     1
## 2      20     2             4 B     1
```

```
## 3      19     3             5 C     1
## 4      18     4             6 D     1
## 5      17     5            21 E     1
## 6      22     6             4 F     2
## 7      22     7            12 G     2
## 8      21     8             8 H     2
## 9      24     9             5 I     0
## 10     24    10             2 J     0
## # ... with 15 more rows
```

2.2.3.2 groupOTU

```
set.seed(2017)
tr <- rtree(4)
x <- as_tibble(tr)
## the input nodes can be node ID or label
groupOTU(x, c('t1', 't4'), group_name = "fake_group")
```

```
## # A tibble: 7 x 5
##   parent  node branch.length label fake_group
##    <int> <int>         <dbl> <chr> <fct>
## 1      5     1       0.435   t4    1
## 2      7     2       0.674   t1    1
## 3      7     3       0.00202 t3    0
## 4      6     4       0.0251  t2    0
## 5      5     5      NA       <NA>  1
## 6      5     6       0.472   <NA>  1
## 7      6     7       0.274   <NA>  1
```

Both `groupClade()` and `groupOTU()` work with the `tbl_tree`, `phylo` and `treedata`, and `ggtree` objects. Here is an example of using `groupOTU()` with a `phylo` tree object.

```
groupOTU(tr, c('t2', 't4'), group_name = "fake_group") %>%
  as_tibble
```

```
## # A tibble: 7 x 5
##   parent  node branch.length label fake_group
##    <int> <int>         <dbl> <chr> <fct>
## 1      5     1       0.435   t4    1
## 2      7     2       0.674   t1    0
## 3      7     3       0.00202 t3    0
## 4      6     4       0.0251  t2    1
## 5      5     5      NA       <NA>  1
## 6      5     6       0.472   <NA>  1
## 7      6     7       0.274   <NA>  0
```

Another example of working with the ggtree object can be found in session 6.4.

The groupOTU will trace back from input nodes to most recent common ancestor. In this example, nodes 1, 4, 5 and 6 are grouping together (4 (t2) -> 6 -> 5 and 1 (t4) -> 5).

Related operational taxonomic units (OTUs) are grouping and they are not necessarily within a clade. They can be monophyletic (clade), polyphyletic or paraphyletic.

```
cls <- list(c1=c("A", "B", "C", "D", "E"),
            c2=c("F", "G", "H"),
            c3=c("L", "K", "I", "J"),
            c4="M")

as_tibble(tree) %>% groupOTU(cls)

## # A tibble: 25 x 5
##     parent  node branch.length label group
##      <int> <int>         <dbl> <chr> <fct>
##  1      20     1             4 A     c1
##  2      20     2             4 B     c1
##  3      19     3             5 C     c1
##  4      18     4             6 D     c1
##  5      17     5            21 E     c1
##  6      22     6             4 F     c2
##  7      22     7            12 G     c2
##  8      21     8             8 H     c2
##  9      24     9             5 I     c3
## 10      24    10             2 J     c3
## # ... with 15 more rows
```

If there are conflicts when tracing back to the most recent common ancestor, users can set overlap parameter to "origin" (the first one counts), "overwrite" (default, the last one counts), or "abandon" (un-selected for grouping)[1].

2.3 Rerooting tree

A phylogenetic tree can be rerooted with a specified outgroup. The **ape** package implements a root() method to reroot a tree stored in a phylo object, while the **treeio** package provides the root() method for treedata object. This method is designed to re-root a phylogenetic tree with associated data concerning the specified outgroup or at the specified node based on the root() implemented in the **ape** package.

We first linked external data to a tree using left_join() and stored all the information in a treedata object, trda.

[1]https://groups.google.com/forum/#!msg/bioc-ggtree/Q4LnwoTf1DM/uqYdYB_VBAAJ

```r
library(ggtree)
library(treeio)
library(tidytree)
library(TDbook)

# load `tree_boots`, `df_tip_data`, and `df_inode_data` from 'TDbook'

trda <- tree_boots %>%
        left_join(df_tip_data, by=c("label" = "Newick_label")) %>%
        left_join(df_inode_data, by=c("label" = "newick_label"))
trda

## 'treedata' S4 object'.
##
## ...@ phylo:
##
## Phylogenetic tree with 7 tips and 6 internal nodes.
##
## Tip labels:
##   Rangifer_tarandus, Cervus_elaphus, Bos_taurus,
## Ovis_orientalis, Suricata_suricatta,
## Cystophora_cristata, ...
## Node labels:
##   Mammalia, Artiodactyla, Cervidae, Bovidae,
## Carnivora, Caniformia
##
## Rooted; includes branch lengths.
##
## with the following features available:
##   '', 'vernacularName', 'imageURL', 'imageLicense',
## 'imageAuthor', 'infoURL', 'mass_in_kg',
## 'trophic_habit', 'ncbi_taxid', 'rank',
## 'vernacularName.y', 'infoURL.y', 'rank.y',
## 'bootstrap', 'posterior'.
##
## # The associated data tibble abstraction: 13 x 17
## # The 'node', 'label' and 'isTip' are from the phylo tree.
##     node label             isTip vernacularName imageURL
##    <int> <chr>             <lgl> <chr>          <chr>
##  1     1 Rangifer_tarand~ TRUE  Reindeer       http://~
##  2     2 Cervus_elaphus   TRUE  Red deer       http://~
##  3     3 Bos_taurus       TRUE  Cattle         https:/~
##  4     4 Ovis_orientalis  TRUE  Asiatic moufl~ http://~
##  5     5 Suricata_surica~ TRUE  Meerkat        http://~
##  6     6 Cystophora_cris~ TRUE  Hooded seal    http://~
```

```
## 7      7 Mephitis_mephit~ TRUE  Striped skunk  http://~
## 8      8 Mammalia         FALSE <NA>           <NA>
## 9      9 Artiodactyla     FALSE <NA>           <NA>
## 10    10 Cervidae         FALSE <NA>           <NA>
## # ... with 3 more rows, and 12 more variables:
## #   imageLicense <chr>, imageAuthor <chr>,
## #   infoURL <chr>, mass_in_kg <dbl>,
## #   trophic_habit <chr>, ncbi_taxid <int>, rank <chr>,
## #   vernacularName.y <chr>, infoURL.y <chr>,
## #   rank.y <chr>, bootstrap <int>, posterior <dbl>
```

Then we can reroot the tree with the associated data mapping to the branches and nodes correctly as demonstrated in Figure 2.2. The figure was visualized using **ggtree** (see also Chapters 4 and 5).

```
# reroot
trda2 <- root(trda, outgroup = "Suricata_suricatta", edgelabel = TRUE)
# The original tree
p1 <- trda %>%
      ggtree() +
      geom_nodelab(
        mapping = aes(
          x = branch,
          label = bootstrap
        ),
        nudge_y = 0.36
      ) +
      xlim(-.1, 4.5) +
      geom_tippoint(
        mapping = aes(
          shape = trophic_habit,
          color = trophic_habit,
          size = mass_in_kg
        )
      ) +
      scale_size_continuous(range = c(3, 10)) +
      geom_tiplab(
        offset = .14,
      ) +
      geom_nodelab(
        mapping = aes(
          label = vernacularName.y,
          fill = posterior
        ),
        geom = "label"
      ) +
```

```
    scale_fill_gradientn(colors = RColorBrewer::brewer.pal(3,
    "YlGnBu")) +
    theme(legend.position = "right")

# after reroot
p2 <- trda2 %>%
    ggtree() +
    geom_nodelab(
      mapping = aes(
        x = branch,
        label = bootstrap
      ),
      nudge_y = 0.36
    ) +
    xlim(-.1, 5.5) +
    geom_tippoint(
      mapping = aes(
        shape = trophic_habit,
        color = trophic_habit,
        size = mass_in_kg
      )
    ) +
    scale_size_continuous(range = c(3, 10)) +
    geom_tiplab(
      offset = .14,
    ) +
    geom_nodelab(
      mapping = aes(
        label = vernacularName.y,
        fill = posterior
      ),
      geom = "label"
    ) +
    scale_fill_gradientn(colors = RColorBrewer::brewer.pal(3,
    "YlGnBu")) +
    theme(legend.position = "right")

plot_list(p1, p2, tag_levels='A', ncol=2)
```

The `outgroup` parameter represents the specific new `outgroup`, it can be a node label (character) or node number. If it is a "single one" value, meaning using the node below this tip as the new root, if it has multiple values, meaning the most recent common of the values will be used as the new root. Note that, if the node labels should be treated as edge labels, the `edgelabel` should be set to `TRUE` to

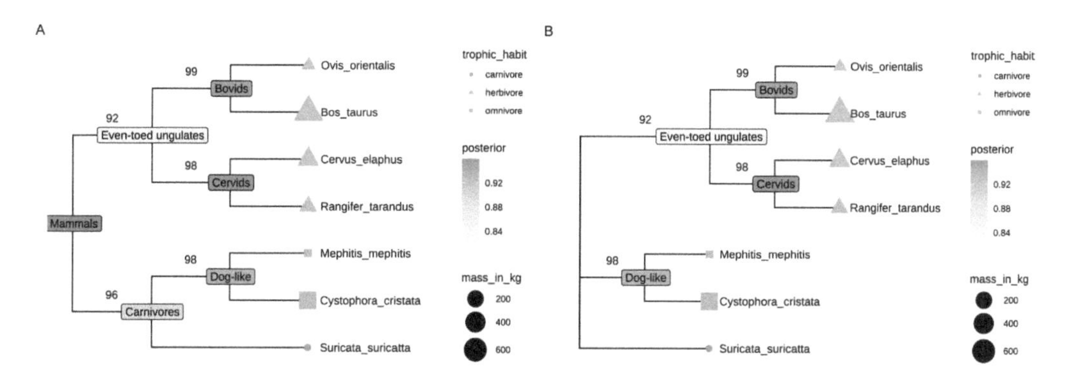

Figure 2.2: **Reroot a phylogenetic tree with associated data.** Original tree (A) and re-rooted tree (B) with associated data mapped to the branches or nodes of the tree correctly. (A) and (B) present before and after rooting on the branch leading to the tip node 'Suricata_suricatta', respectively.

return the correct relationship between the `node` and `associated data`. For more details about re-root, including precautions and pitfalls, please refer to the review article (Czech et al., 2017).

2.4 Rescaling Tree Branches

Phylogenetic data can be merged for joint analysis (Figure 2.1). They can be displayed on the same tree structure as a more complex annotation to help visually inspection of their evolutionary patterns. All the numerical data stored in a `treedata` object can be used to re-scale tree branches. For example, CodeML infers d_N/d_S, d_N, and d_S, all these statistics can be used as branch lengths (Figure 2.3). All these values can also be used to color the tree (session 4.3.4) and can be projected to a vertical dimension to create a two-dimensional tree or phenogram (session 4.2.2 and Figures 4.5 and 4.11).

```
p1 <- ggtree(merged_tree) + theme_tree2()
p2 <- ggtree(rescale_tree(merged_tree, 'dN')) + theme_tree2()
p3 <- ggtree(rescale_tree(merged_tree, 'rate')) + theme_tree2()

plot_list(p1, p2, p3, ncol=3, tag_levels='A')
```

Modifying branch lengths in the tree object in addtion to using the `rescale_tree()` function, users can directly specify a variable as branch length in `ggtree()` as demonstrated in session 4.3.6.

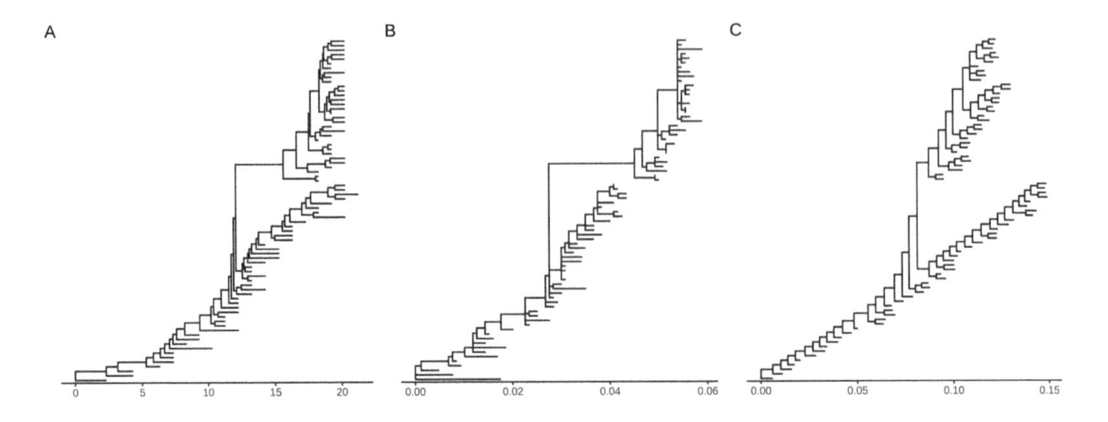

Figure 2.3: **Rescaling tree branches.** The tree with branches scaled in time (year from the root) (A). The tree was rescaled using d_N as branch lengths (B). The tree was rescaled using substitution rates (C).

2.5 Subsetting Tree with Data

2.5.1 Removing tips in a phylogenetic tree

Sometimes we want to remove selected tips from a phylogenetic tree. This is due to several reasons, including low sequence quality, errors in sequence assembly, an alignment error in part of the sequence, an error in phylogenetic inference, *etc.*

Let's say that we want to remove three tips (colored red) from the tree (Figure 2.4A), the `drop.tip()` method removes specified tips and updates the tree (Figure 2.4B). All associated data will be maintained in the updated tree.

```
f <- system.file("extdata/NHX", "phyldog.nhx", package="treeio")
nhx <- read.nhx(f)
to_drop <- c("Physonect_sp_@2066767",
             "Lychnagalma_utricularia@2253871",
             "Kephyes_ovata@2606431")
p1 <- ggtree(nhx) + geom_tiplab(aes(color = label %in% to_drop)) +
  scale_color_manual(values=c("black", "red")) + xlim(0, 0.8)

nhx_reduced <- drop.tip(nhx, to_drop)
p2 <- ggtree(nhx_reduced) + geom_tiplab() + xlim(0, 0.8)
plot_list(p1, p2, ncol=2, tag_levels = "A")
```

2.5.2 Subsetting tree by tip label

Sometimes a tree can be large and difficult to look at only the portions of interest. The `tree_subset()` function was created in the **treeio** package (Wang et al., 2020) to extract a subset of the tree portion while still maintaining the structure of the tree portion. The `beast_tree` in Figure 2.5A is slightly crowded. Obviously, we

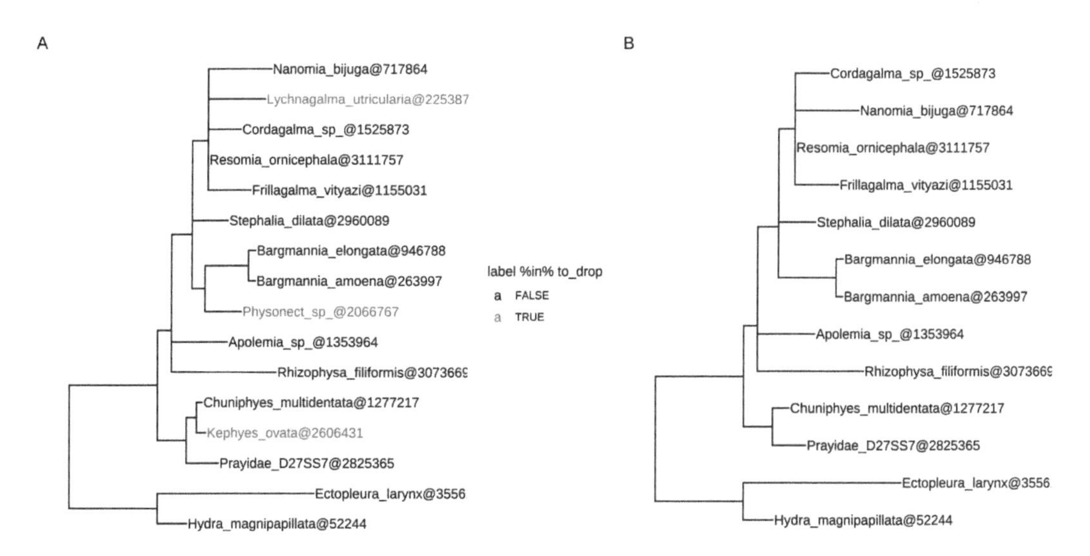

Figure 2.4: **Removing tips from a tree.** Original tree with three tips (colored red) to remove (A). The updated tree removed selected tips (B).

can make the figure taller to allow more space for the labels (similar to using the "Expansion" slider in `FigTree`) or we can make the text smaller. However, these solutions are not always applicable when you have a lot of tips (*e.g.*, hundreds or thousands of tips). In particular, when you are only interested in the portion of the tree around a particular tip, you certainly don't want to explore a large tree to find the certain species you are interested in.

Let's say you are interested in tip *A/Swine/HK/168/2012* from the tree (Figure 2.5A) and you want to look at the immediate relatives of this tip.

The `tree_subset()` function allows you to look at the portions of the tree that are of interest. By default, the `tree_subset()` function will internally call the `groupOTU()` to assign the group specified tip from the rest of the other tips (Figure 2.5B). Additionally, the branch lengths and related associated data are maintained after subsetting (Figure 2.5C). The root of the tree is always anchored at zero for the subset tree by default and all the distances are relative to this root. If you want all the distances to be relative to the original root, you can specify the root position (by `root.position` parameter) to the root edge of the subset tree, which is the sum of branch lengths from the original root to the root of the subset tree (Figures 2.5D and E).

```
beast_file <- system.file("examples/MCC_FluA_H3.tree",
package="ggtree")
beast_tree <- read.beast(beast_file)

p1 = ggtree(beast_tree) +
  geom_tiplab(offset=.05) +  xlim(0, 40) + theme_tree2()
```

```
tree2 = tree_subset(beast_tree, "A/Swine/HK/168/2012", levels_back=4)
p2 <- ggtree(tree2, aes(color=group)) +
  scale_color_manual(values = c("black", "red"), guide = 'none') +
  geom_tiplab(offset=.2) +  xlim(0, 4.5) + theme_tree2()

p3 <- p2 +
  geom_point(aes(fill = rate), shape = 21, size = 4) +
  scale_fill_continuous(low = 'blue', high = 'red') +
  xlim(0,5) + theme(legend.position = 'right')

p4 <- ggtree(tree2, aes(color=group),
          root.position = as.phylo(tree2)$root.edge) +
  geom_tiplab() + xlim(18, 24) +
  scale_color_manual(values = c("black", "red"), guide = 'none') +
  theme_tree2()

p5 <- p4 +
  geom_rootedge() + xlim(0, 50)

plot_list(p1, p2, p3, p4, p5,
        design="AABBCC\nAADDEE", tag_levels='A')
```

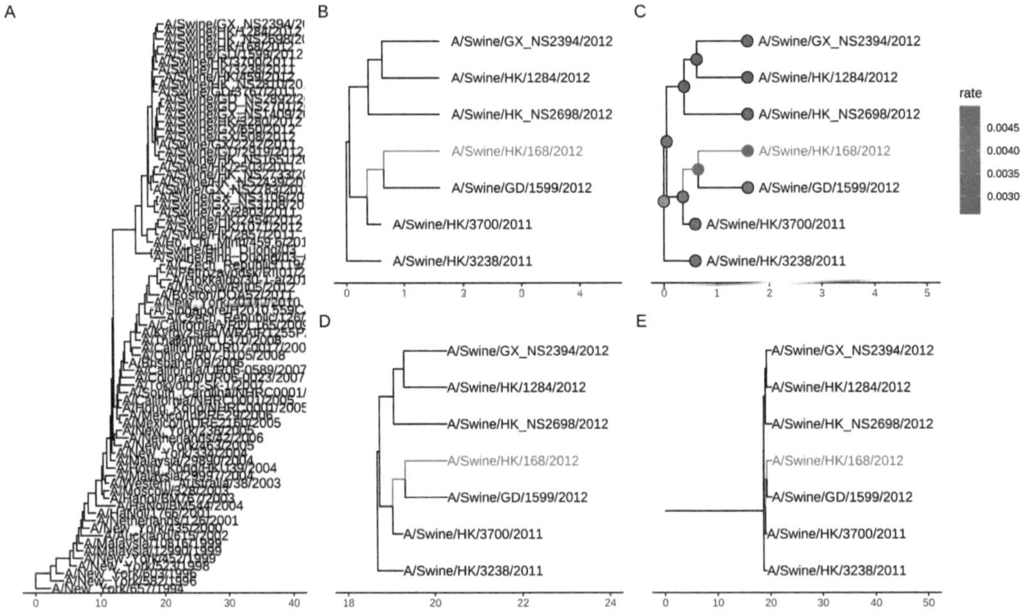

Figure 2.5: **Subsetting tree for a specific tip.** Original tree (A). Subset tree (B). Subset tree with data (C). Visualize the subset tree relative to the original position, without rootedge (D) and with rootedge (E).

2.5.3 Subsetting tree by internal node number

If you are interested in a certain clade, you can specify the input node as an internal node number. The `tree_subset()` function will take the clade as a whole and also trace it back to particular levels to look at the immediate relatives of the clade (Figures 2.6A and B). We can use the `tree_subset()` function to zoom in selected portions and plot a whole tree with the portion of it, which is similar to the `ape::zoom()` function to explore a very large tree (Figures 2.6C and D). Users can also use `viewClade()` function to restrict tree visualization at specific clade as demonstrated in session 6.1.

```
clade <- tree_subset(beast_tree, node=121, levels_back=0)
clade2 <- tree_subset(beast_tree, node=121, levels_back=2)
p1 <- ggtree(clade) + geom_tiplab() + xlim(0, 5)
p2 <- ggtree(clade2, aes(color=group)) + geom_tiplab() +
  xlim(0, 9) + scale_color_manual(values=c("black", "red"))


library(ape)
library(tidytree)
library(treeio)

data(chiroptera)

nodes <- grep("Plecotus", chiroptera$tip.label)
chiroptera <- groupOTU(chiroptera, nodes)

clade <- MRCA(chiroptera, nodes)
x <- tree_subset(chiroptera, clade, levels_back = 0)

p3 <- ggtree(chiroptera, aes(colour = group)) +
  scale_color_manual(values=c("black", "red")) +
  theme(legend.position = "none")
p4 <- ggtree(x) + geom_tiplab() + xlim(0, 6)
plot_list(p1, p2, p3, p4,
  ncol=2, tag_levels = 'A')
```

2.6 Manipulating Tree Data for Visualization

Tree visualization is supported by **ggtree** (Yu et al., 2017). Although **ggtree** implemented several methods for visual exploration of trees with data, you may want to do something that is not supported directly. In this case, you need to manipulate tree data with node coordination positions that are used for visualization. This is quite easy with **ggtree**. Users can use the `fortify()` method which internally calls `tidytree::as_tibble()` to convert the tree to a tidy data frame and add columns

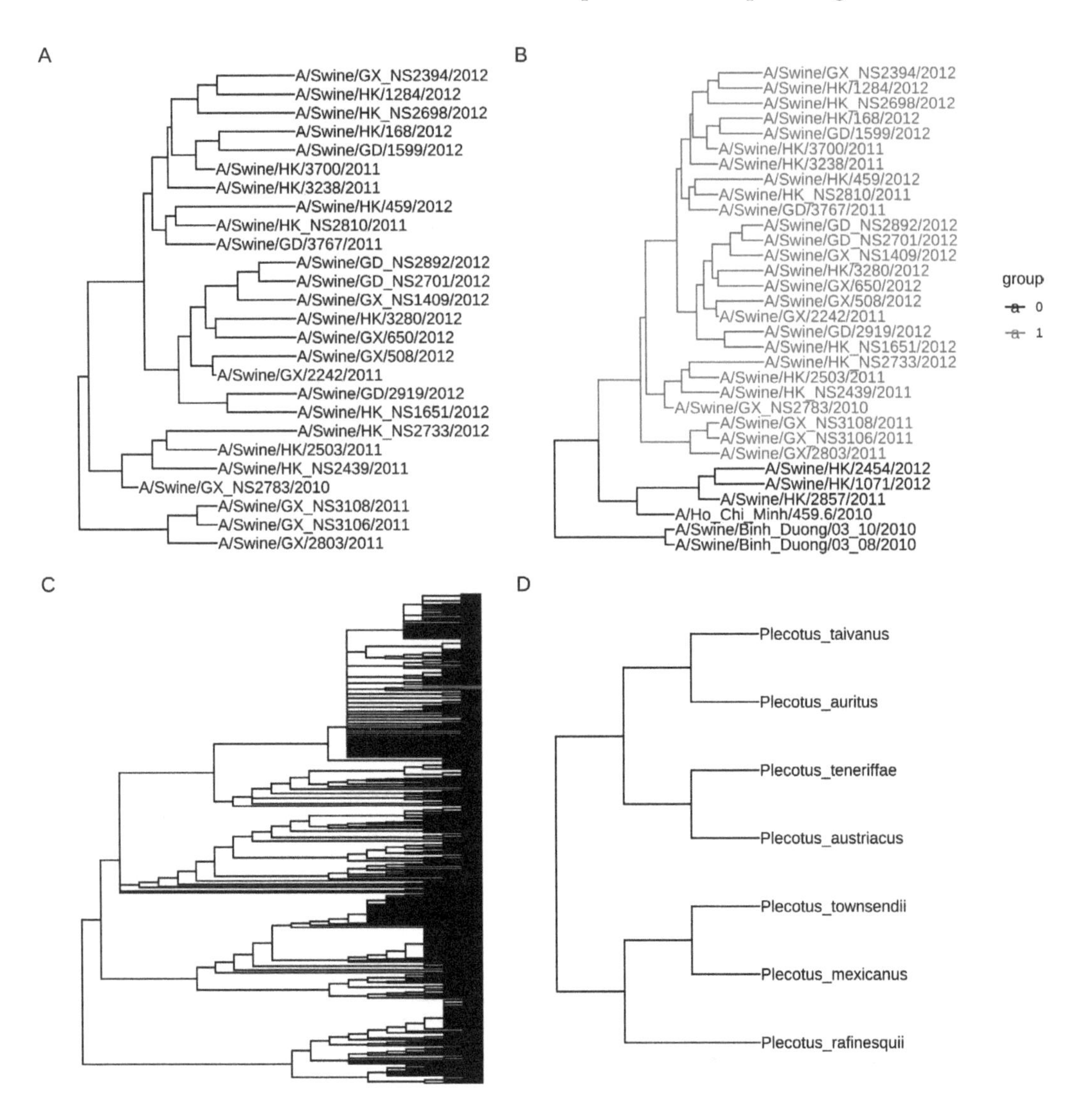

Figure 2.6: **Subsetting tree for the specific clade.** Extracting a clade (A). Extracting a clade and tracing it back to look at its immediate relatives (B). Viewing a very large tree (C) and a selected portion of it (D).

of coordination positions (*i.e.*, x, y, branch, and angle) that are used to plot the tree. You can also access the data via `ggtree(tree)$data`.

Here is an example to plot two trees face-to-face that is similar to a graph produced by the `ape::cophyloplot()` function (Figure 2.7).

```
library(dplyr)
library(ggtree)

set.seed(1024)
```

```
x <- rtree(30)
y <- rtree(30)
p1 <- ggtree(x, layout='roundrect') +
  geom_hilight(
         mapping=aes(subset = node %in% c(38, 48, 58, 36),
                     node = node,
                     fill = as.factor(node)
                     )
     ) +
    labs(fill = "clades for tree in left" )

p2 <- ggtree(y)

d1 <- p1$data
d2 <- p2$data

## reverse x-axis and
## set offset to make the tree on the right-hand side of
## the first tree
d2$x <- max(d2$x) - d2$x + max(d1$x) + 1

pp <- p1 + geom_tree(data=d2, layout='ellipse') +
  ggnewscale::new_scale_fill() +
  geom_hilight(
         data = d2,
         mapping = aes(
            subset = node %in% c(38, 48, 58),
            node=node,
            fill=as.factor(node))
  ) +
  labs(fill = "clades for tree in right" )

dd <- bind_rows(d1, d2) %>%
  filter(!is.na(label))

pp + geom_line(aes(x, y, group=label), data=dd, color='grey') +
    geom_tiplab(geom = 'shadowtext', bg.colour = alpha('firebrick',
    .5)) +
    geom_tiplab(data = d2, hjust=1, geom = 'shadowtext',
                bg.colour = alpha('firebrick', .5))
```

It is quite easy to plot multiple trees and connect taxa in one figure; for instance, plotting trees constructed from all internal gene segments of influenza virus and connecting equivalent strains across the trees (Venkatesh et al., 2018). Figure 2.8

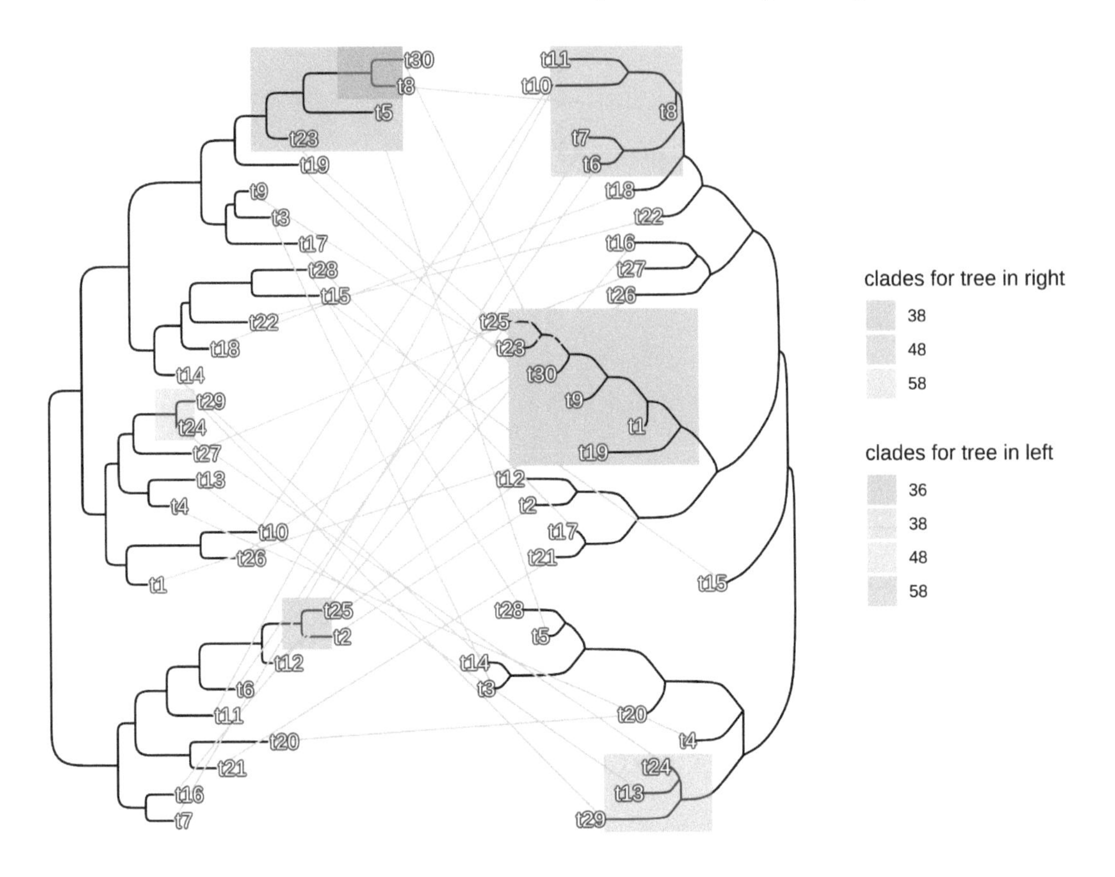

Figure 2.7: **Plot two phylogenetic trees face-to-face.** Plotting a tree using `ggtree()` (left-hand side) and subsequently adding another layer of a tree by `geom_tree()` (right-hand side). The relative positions of the plotted trees can be manually adjusted and adding layers to each of the trees (*e.g.*, tip labels and highlighting clades) is independent.

demonstrates the usage of plotting multiple trees by combining multiple layers of `geom_tree()`.

```
z <- rtree(30)
d2 <- fortify(y)
d3 <- fortify(z)
d2$x <- d2$x + max(d1$x) + 1
d3$x <- d3$x + max(d2$x) + 1

dd = bind_rows(d1, d2, d3) %>%
  filter(!is.na(label))

p1 + geom_tree(data = d2) + geom_tree(data = d3) + geom_tiplab(data=d3) +
  geom_line(aes(x, y, group=label, color=node < 15), data=dd, alpha=.3)
```

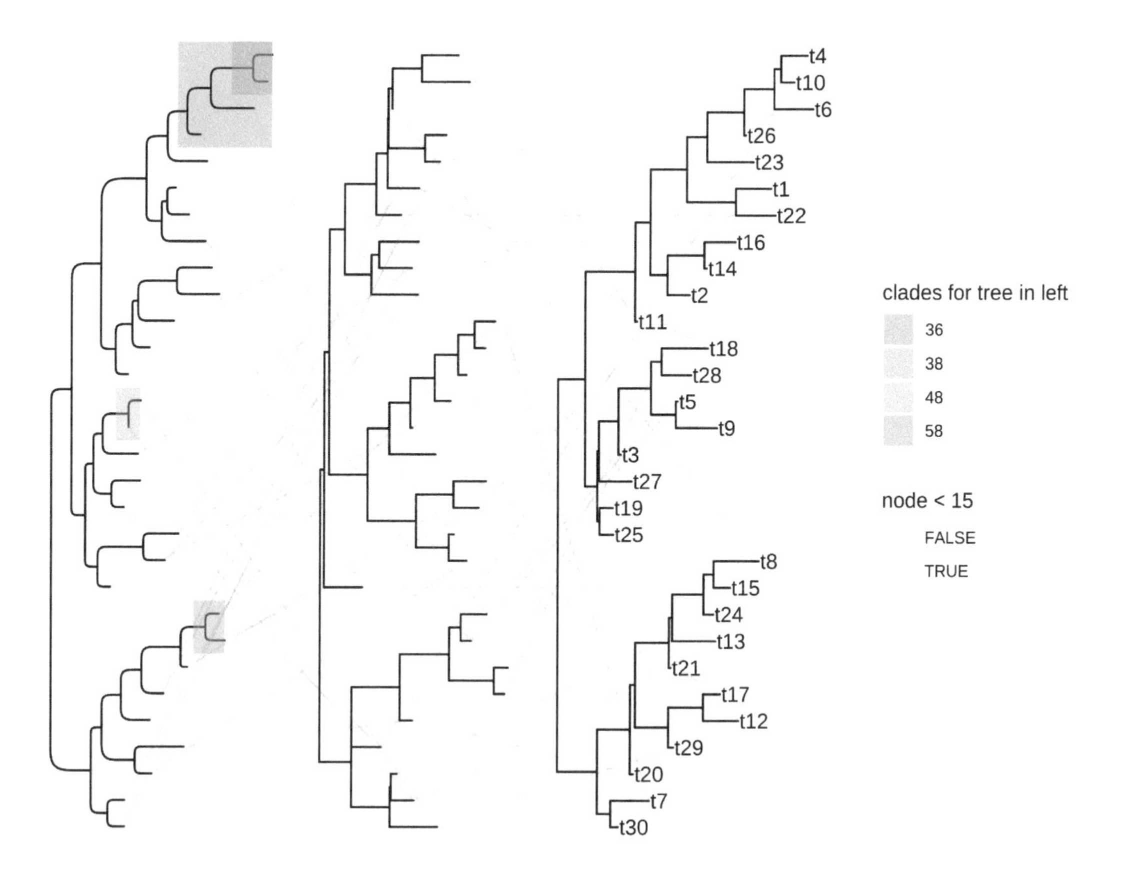

Figure 2.8: **Plot multiple phylogenetic trees side-by-side.** Plotting a tree using `ggtree()` and subsequently adding multiple layers of trees by `geom_tree()`.

2.7 Summary

The **treeio** package allows us to import diverse phylogeny associated data into R. However, a phylogenetic tree is stored in a way to facilitate computational processing which is not human friendly and needs the expertise to manipulate and explore tree data. The **tidytree** package provides a tidy interface for exploring tree data, while **ggtree** provides a set of utilities to visualize and explore tree data using the grammar of graphics. This full suite of packages makes it easy for ordinary users to interact with tree data and allows us to integrate phylogeny associated data from different sources (*e.g.*, experimental results or analysis findings), which creates the possibilities of integrative and comparative study. Moreover, this package suite brings phylogenetic analysis into the tidyverse and certainly takes us to the next level of processing phylogenetic data.

Chapter 3

Exporting Tree with Data

3.1 Introduction

The **treeio** package (Wang et al., 2020) supports parsing various phylogenetic tree file formats including software outputs that contain evolutionary evidence. Some of the formats are just log files (*e.g.*, **PAML** and **r8s** output), while some of the others are non-standard formats (*e.g.*, **BEAST** and **MrBayes** output that introduce a square bracket, which was reserved to store comments in standard Nexus format, to store inferences). With **treeio**, we are now able to parse these files to extract phylogenetic trees and map associated data on the tree structure. Exporting tree structure is easy, users can use the `as.phyo()` method defined in **treeio** to convert a `treedata` object to a `phylo` object and then use `write.tree()` or `write.nexus()` implemented in **ape** package (Paradis et al., 2004) to export the tree structure as Newick text or Nexus file. This is quite useful for converting non-standard formats to a standard format and for extracting trees from software outputs, such as log files.

However, exporting a tree with associated data is still challenging. These associated data can be parsed from analysis programs or obtained from external sources (*e.g.*, phenotypic data, experimental data, and clinical data). The major obstacle here is that there is no standard format designed for storing a tree with data. NeXML (Vos et al., 2012) may be the most flexible format. However, it is currently not widely supported. Most of the analysis programs in this field rely extensively on Newick string and Nexus format. In my opinion, although BEAST Nexus format may not be the best solution, it is currently a good approach for storing heterogeneous associated data. The beauty of the format is that all the annotated elements are stored within square brackets, which are reserved for comments. In this way, existing programs that can read standard Nexus format are able to parse it by ignoring the annotated elements.

3.2 Exporting Tree Data to *BEAST* Nexus Format

3.2.1 Exporting/converting software output

The **treeio** package (Wang et al., 2020) provides the `write.beast()` function to export `treedata` object as BEAST Nexus file (Bouckaert et al., 2014). With **treeio**, it is easy to convert software output to BEAST format if the output can be parsed by **treeio** (see Chapter 1).

Here is an example of converting NHX file to BEAST format:

```
nhxfile <- system.file("extdata/NHX", "phyldog.nhx", package="treeio")
nhx <- read.nhx(nhxfile)
# write.beast(nhx, file = "phyldog.tree")
write.beast(nhx)
```

```
#NEXUS
[R-package treeio, Thu Oct 14 11:24:19 2021]

BEGIN TAXA;
    DIMENSIONS NTAX = 16;
    TAXLABELS
        Prayidae_D27SS7@2825365
        Kephyes_ovata@2606431
        Chuniphyes_multidentata@1277217
        Apolemia_sp_@1353964
        Bargmannia_amoena@263997
        Bargmannia_elongata@946788
        Physonect_sp_@2066767
        Stephalia_dilata@2960089
        Frillagalma_vityazi@1155031
        Resomia_ornicephala@3111757
        Lychnagalma_utricularia@2253871
        Nanomia_bijuga@717864
        Cordagalma_sp_@1525873
        Rhizophysa_filiformis@3073669
        Hydra_magnipapillata@52244
        Ectopleura_larynx@3556167
    ;
END;
BEGIN TREES;
    TRANSLATE
        1   Prayidae_D27SS7@2825365,
        2   Kephyes_ovata@2606431,
        3   Chuniphyes_multidentata@1277217,
        4   Apolemia_sp_@1353964,
        5   Bargmannia_amoena@263997,
        6   Bargmannia_elongata@946788,
        7   Physonect_sp_@2066767,
        8   Stephalia_dilata@2960089,
        9   Frillagalma_vityazi@1155031,
```

```
        10  Resomia_ornicephala@3111757,
        11  Lychnagalma_utricularia@2253871,
        12  Nanomia_bijuga@717864,
        13  Cordagalma_sp_@1525873,
        14  Rhizophysa_filiformis@3073669,
        15  Hydra_magnipapillata@52244,
        16  Ectopleura_larynx@3556167
    ;
    TREE * UNTITLED = [&R] (((1[&Ev=S,ND=0,S=58]:0.0682841,(2[&Ev=S,ND=1,
S=69]:0.0193941,3[&Ev=S,ND=2,S=70]:0.0121378)[&Ev=S,ND=3,S=60]:0.0217782)
[&Ev=S,ND=4,S=36]:0.0607598,((4[&Ev=S,ND=9,S=31]:0.11832,(((5[&Ev=S,ND=10,
S=37]:0.0144549,6[&Ev=S,ND=11,S=38]:0.0149723)[&Ev=S,ND=12,S=33]:0.0925388,
7[&Ev=S,ND=13,S=61]:0.077429)[&Ev=S,ND=14,S=24]:0.0274637,(8[&Ev=S,ND=15,
S=52]:0.0761163,((9[&Ev=S,ND=16,S=53]:0.0906068,10[&Ev=S,ND=17,S=54]:1e-06)
[&Ev=S,ND=18,S=45]:1e-06,((11[&Ev=S,ND=19,S=65]:0.120851,12[&Ev=S,ND=20,
S=71]:0.133939)[&Ev=S,ND=21,S=56]:1e-06,13[&Ev=S,ND=22,S=64]:0.0693814)
[&Ev=S,ND=23,S=46]:1e-06)[&Ev=S,ND=24,S=40]:0.0333823)[&Ev=S,ND=25,S=35]:
1e-06)[&Ev=D,ND=26,S=24]:0.0431861)[&Ev=S,ND=27,S=19]:1e-06,14[&Ev=S,ND=28,
S=26]:0.22283)[&Ev=S,ND=29,S=17]:0.0292362)[&Ev=D,ND=8,S=17]:0.185603,
(15[&Ev=S,ND=5,S=16]:0.0621782,16[&Ev=S,ND=6,S=15]:0.332505)[&Ev=S,ND=7,
S=12]:0.185603)[&Ev=S,ND=30,S=9];
END;
```

Another example of converting **CODEML** output to BEAST format:

```
mlcfile <- system.file("extdata/PAML_Codeml", "mlc", package="treeio")
ml <- read.codeml_mlc(mlcfile)
# write.beast(ml, file = "codeml.tree")
write.beast(ml) # output not shown
```

Some software tools that do not support these outputs can be supported through data conversion. For example, we can convert the NHX file to BEAST file and use NHX tags to color the tree using **FigTree** (Figure 3.1A) or convert **CODEML** output and use d_N/d_S, d_N, or d_S to color the tree in **FigTree** (Figure 3.1B). Before conversion, these files could not be opened in Figtree. Treeio's conversion function makes data available to other software tools and expands the application range of these tools.

3.2.2 Combining tree with external data

Using the utilities provided by **tidytree** and **treeio**, it is easy to link external data onto the corresponding phylogeny. The `write.beast()` function enables users to export the tree with external data to a single tree file.

```
phylo <- as.phylo(nhx)
## save space for printing the tree text
phylo$edge.length <- round(phylo$edge.length, 2)

## print the newick text
```

A

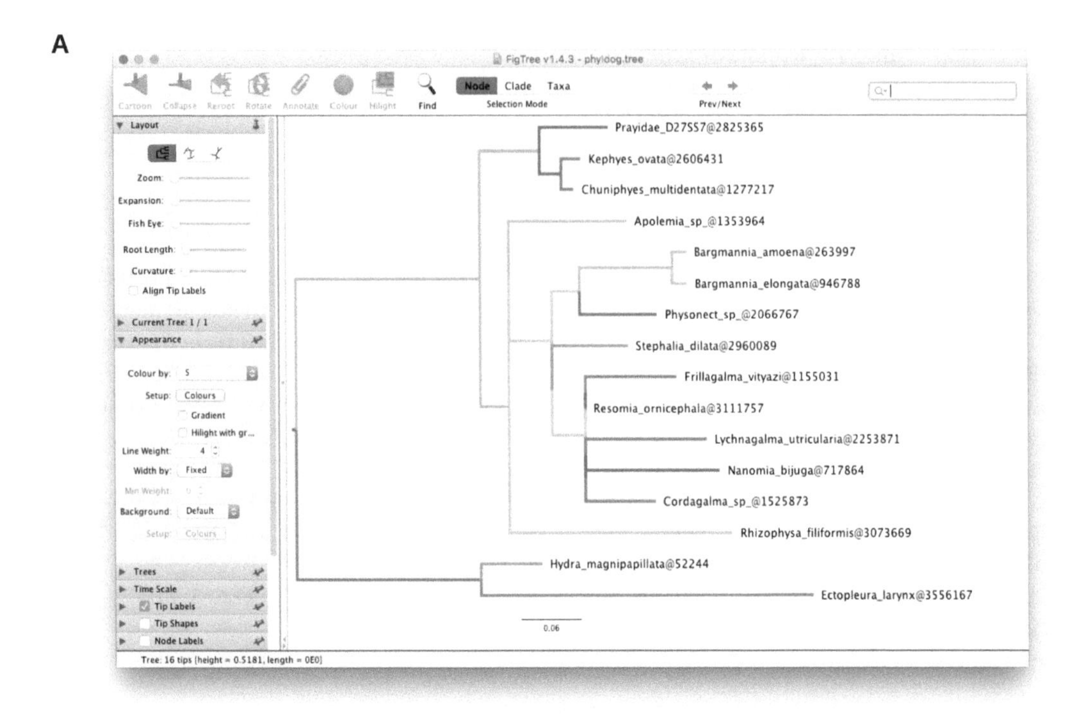

B

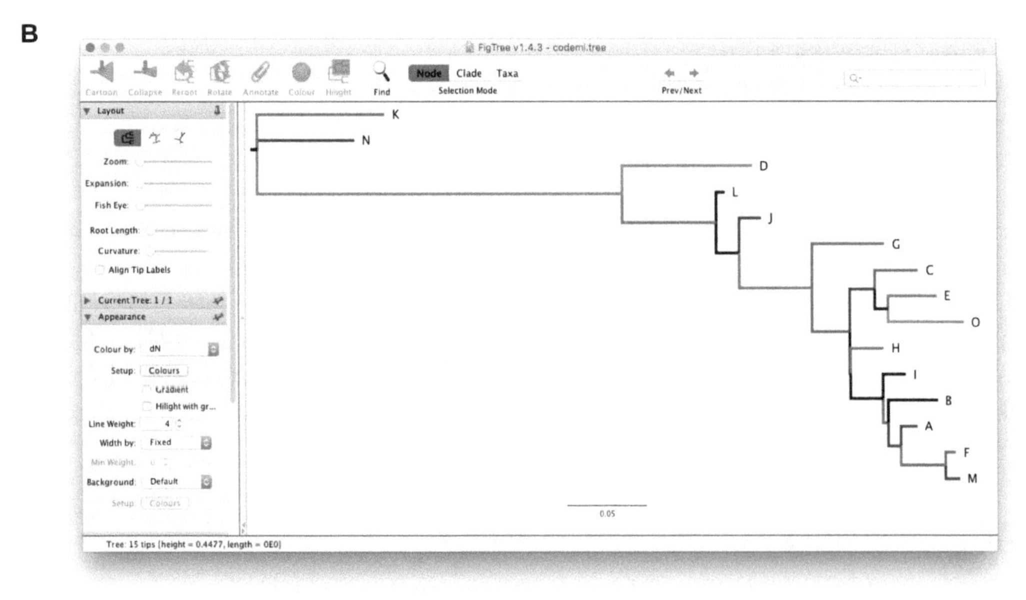

Figure 3.1: **Visualizing BEAST file in FigTree.** Directly visualizing `NHX` file (A) and `CodeML` output (B) in `FigTree` is not supported. `treeio` can convert these files to BEAST compatible NEXUS format which can be directly opened in `FigTree` and visualized together with annotated data.

```
write.tree(phylo)
```

```
(((Prayidae_D27SS7@2825365:0.07,(Kephyes_ovata@2606431:0.02,
Chuniphyes_multidentata@1277217:0.01):0.02):0.06,((Apolemia_sp_@1353964:0.12,
(((Bargmannia_amoena@263997:0.01,Bargmannia_elongata@946788:0.01):0.09,
Physonect_sp_@2066767:0.08):0.03,(Stephalia_dilata@2960089:0.08,
((Frillagalma_vityazi@1155031:0.09,Resomia_ornicephala@3111757:0):0,
((Lychnagalma_utricularia@2253871:0.12,Nanomia_bijuga@717864:0.13):0,
Cordagalma_sp_@1525873:0.07):0):0.03):0):0.04):0,
Rhizophysa_filiformis@3073669:
0.22):0.03):0.19,(Hydra_magnipapillata@52244:0.06,
Ectopleura_larynx@3556167:0.33):0.19);
```

```
N <- Nnode2(phylo)
fake_data <- tibble(node = 1:N, fake_trait = round(rnorm(N), 2),
                    another_trait = round(runif(N), 2))
fake_tree <- full_join(phylo, fake_data, by = "node")
# write.beast(fake_tree)

## to save space, use a subtree
fake_tree2 = tree_subset(fake_tree, node=27, levels_back=0)
write.beast(fake_tree2)
```

```
#NEXUS
[R-package treeio, Tue Nov 16 10:13:32 2021]

BEGIN TAXA;
    DIMENSIONS NTAX = 5;
    TAXLABELS
        Frillagalma_vityazi@1155031
        Resomia_ornicephala@3111757
        Lychnagalma_utricularia@2253871
        Nanomia_bijuga@717864
        Cordagalma_sp_@1525873
    ;
END;
BEGIN TREES;
    TRANSLATE
        1   Frillagalma_vityazi@1155031,
        2   Resomia_ornicephala@3111757,
        3   Lychnagalma_utricularia@2253871,
        4   Nanomia_bijuga@717864,
        5   Cordagalma_sp_@1525873
    ;
    TREE * UNTITLED = [&R] ((1[&fake_trait=-1.42,another_trait=0.17]:0.09,
2[&fake_trait=0.19,another_trait=0.04]:0)[&fake_trait=0.85,
another_trait=0.56]:0,(5[&fake_trait=0.22,another_trait=0.73]:0.07,
(3[&fake_trait=0.02,another_trait=0.29]:0.12,4[&fake_trait=-1.29,
another_trait=0.35]:0.13)[&fake_trait=-0.33,another_trait=0.88]:0)
[&fake_trait=0.27,another_trait=0.94]:0):0.29;
```

```
END;
```

After merging, the `fake_trait` and `another_trait` stored in `fake_data` will be linked to the tree, `phylo`, and stored in the `treedata` object, the `fake_tree`. The `write.beast()` function exports the tree with associated data to a single BEAST format file. The associated data can be used to visualize the tree using **ggtree** (Figure 5.8) or **FigTree** (Figure 3.1).

3.2.3 Merging tree data from different sources

Not only Newick tree text can be combined with associated data, but also tree data obtained from software output can be combined with external data, as well as different tree objects can be merged (for details, see Chapter 2).

```
## combine tree object with data
tree_with_data <- full_join(nhx, fake_data, by = "node")
tree_with_data
```

```
## 'treedata' S4 object that stored information
## of
##  '/home/ygc/R/library/treeio/extdata/NHX/phyldog.nhx'.
##
## ...@ phylo:
##
## Phylogenetic tree with 16 tips and 15 internal nodes.
##
## Tip labels:
##   Prayidae_D27SS7@2825365, Kephyes_ovata@2606431,
## Chuniphyes_multidentata@1277217,
## Apolemia_sp_@1353964, Bargmannia_amoena@263997,
## Bargmannia_elongata@946788, ...
##
## Rooted; includes branch lengths.
##
## with the following features available:
##   'Ev', 'ND', 'S', 'fake_trait', 'another_trait'.
##
## # The associated data tibble abstraction: 31 x 8
## # The 'node', 'label' and 'isTip' are from the phylo tree.
##      node label       isTip Ev       ND     S fake_trait
##     <int> <chr>       <lgl> <chr> <dbl> <dbl>      <dbl>
##  1      1 Prayidae_D~ TRUE  S         0    58       0.81
##  2      2 Kephyes_ov~ TRUE  S         1    69       1.05
##  3      3 Chuniphyes~ TRUE  S         2    70      -0.53
##  4      4 Apolemia_s~ TRUE  S         9    31       1.26
##  5      5 Bargmannia~ TRUE  S        10    37       1.26
```

```
## 6      6 Bargmannia~ TRUE  S          11     38         0.24
## 7      7 Physonect_~ TRUE  S          13     61         0.73
## 8      8 Stephalia_~ TRUE  S          15     52         1.34
## 9      9 Frillagalm~ TRUE  S          16     53        -0.79
## 10    10 Resomia_or~ TRUE  S          17     54         0.15
## # ... with 21 more rows, and 1 more variable:
## #   another_trait <dbl>
```

```
## merge two tree object
tree2 <- merge_tree(nhx, fake_tree)
identical(tree_with_data, tree2)
```

```
## [1] TRUE
```

After merging data from different sources, the tree with the associated data can be exported into a single file.

```
outfile <- tempfile(fileext = ".tree")
write.beast(tree2, file = outfile)
```

The output BEAST Nexus file can be imported into R using the `read.beast` function and all the associated data can be used to annotate the tree using ggtree (Yu et al., 2017).

```
read.beast(outfile)
```

```
## 'treedata' S4 object that stored information
## of
##  '/tmp/RtmpFUmFLn/file24fe918edab90.tree'.
##
## ...@ phylo:
##
## Phylogenetic tree with 16 tips and 15 internal nodes.
##
## Tip labels:
##   Prayidae_D27SS7@2825365, Kephyes_ovata@2606431,
## Chuniphyes_multidentata@1277217,
## Apolemia_sp_@1353964, Bargmannia_amoena@263997,
## Bargmannia_elongata@946788, ...
##
## Rooted; includes branch lengths.
##
## with the following features available:
##   'another_trait', 'Ev', 'fake_trait', 'ND', 'S'.
##
## # The associated data tibble abstraction: 31 x 8
## # The 'node', 'label' and 'isTip' are from the phylo tree.
##     node label     isTip another_trait Ev    fake_trait
```

```
##     <int> <chr>     <lgl> <chr>          <chr> <chr>
##  1      1 Prayidae~ TRUE  0.1            S     0.81
##  2      2 Kephyes_~ TRUE  0.26           S     1.05
##  3      3 Chuniphy~ TRUE  0.08           S     -0.53
##  4      4 Apolemia~ TRUE  0.29           S     1.26
##  5      5 Bargmann~ TRUE  0.88           S     1.26
##  6      6 Bargmann~ TRUE  0.38           S     0.24
##  7      7 Physonec~ TRUE  0.5            S     0.73
##  8      8 Stephali~ TRUE  0.79           S     1.34
##  9      9 Frillaga~ TRUE  0.44           S     -0.79
## 10     10 Resomia_~ TRUE  0.1            S     0.15
## # ... with 21 more rows, and 2 more variables:
## #   ND <chr>, S <chr>
```

3.3 Exporting Tree Data to the *jtree* Format

The **treeio** package (Wang et al., 2020) provides the `write.beast()` function to export `treedata` to BEAST Nexus file. This is quite useful to convert file format, combine tree with data and merge tree data from different sources as we demonstrated in the previous session. The **treeio** package also supplies the `read.beast()` function to parse the output file of the `write.beast()` function. Although with **treeio**, the R community has the ability to manipulate BEAST Nexus format and process tree data, there is still a lacking library/package for parsing BEAST files in other programming languages.

JSON (JavaScript Object Notation) is a lightweight data-interchange format and is widely supported in almost all modern programming languages. To make it easy to import a tree with data in other programming languages, **treeio** supports exporting a tree with data in the `jtree` format, which is JSON-based and can be easy to parse using any language that supports JSON.

```
# write.jtree(tree2)

# to save space, use a subtree
tree3 <- tree_subset(tree2, node=24, levels_back=0)
write.jtree(tree3)
```


```
{
    "tree": "(Physonect_sp_@2066767:0.077429{3},(Bargmannia_amoena@263997:
    0.0144549
{1},Bargmannia_elongata@946788:0.0149723{2}):0.0925388{5}):0.28549{4};",
    "data":[
  {
    "edge_num": 1,
    "Ev": "S",
    "ND": 10,
    "S": 37,
```

```
    "fake_trait": -0.69,
    "another_trait": 0.42
  },
  {
    "edge_num": 2,
    "Ev": "S",
    "ND": 11,
    "S": 38,
    "fake_trait": -0.95,
    "another_trait": 0.38
  },
  {
    "edge_num": 3,
    "Ev": "S",
    "ND": 13,
    "S": 61,
    "fake_trait": 0.59,
    "another_trait": 0.65
  },
  {
    "edge_num": 4,
    "Ev": "S",
    "ND": 14,
    "S": 24,
    "fake_trait": -0.69,
    "another_trait": 0.06
  },
  {
    "edge_num": 5,
    "Ev": "S",
    "ND": 12,
    "S": 33,
    "fake_trait": -0.58,
    "another_trait": 0.4
  }
],
    "metadata": {"info": "R-package treeio", "data":
    "Tue Nov 16 10:21:20 2021"}
}
```


The `jtree` format is based on JSON and can be parsed using JSON parser.

```
jtree_file <- tempfile(fileext = '.jtree')
write.jtree(tree3, file = jtree_file)
jsonlite::fromJSON(jtree_file)
```

```
$tree
[1] "(Physonect_sp_@2066767:0.077429{3},(Bargmannia_amoena
@263997:0.0144549{1},
Bargmannia_elongata@946788:0.0149723{2}):0.0925388{5}):0.28549{4};"
```

```
$data
  edge_num Ev ND  S fake_trait another_trait
1        1  S 10 37      -0.69          0.42
2        2  S 11 38      -0.95          0.38
3        3  S 13 61       0.59          0.65
4        4  S 14 24      -0.69          0.06
5        5  S 12 33      -0.58          0.40

$metadata
$metadata$info
[1] "R-package treeio"

$metadata$data
[1] "Tue Nov 16 10:24:34 2021"
```

The `jtree` file can be directly imported as a `treedata` object using the `read.jtree()` function provided also in **treeio** package (see also session 1.3).

```
read.jtree(jtree_file)

## 'treedata' S4 object that stored information
## of
##  '/tmp/RtmpFUmFLn/file24fe93421308a.jtree'.
##
## ...@ phylo:
##
## Phylogenetic tree with 3 tips and 2 internal nodes.
##
## Tip labels:
##   Physonect_sp_@2066767, Bargmannia_amoena@263997,
## Bargmannia_elongata@946788
##
## Rooted; includes branch lengths.
##
## with the following features available:
##   'Ev', 'ND', 'S', 'fake_trait', 'another_trait'.
##
## # The associated data tibble abstraction: 5 x 8
## # The 'node', 'label' and 'isTip' are from the phylo tree.
##    node label         isTip Ev       ND     S fake_trait
##   <int> <chr>         <lgl> <chr> <int> <int>      <dbl>
## 1     1 Physonect_s~ TRUE  S        13    61       0.73
## 2     2 Bargmannia_~ TRUE  S        10    37       1.26
## 3     3 Bargmannia_~ TRUE  S        11    38       0.24
## 4     4 <NA>         FALSE S        14    24       0.87
```

```
## 5      5 <NA>          FALSE S          12    33         2.56
## # ... with 1 more variable: another_trait <dbl>
```

3.4 Summary

Phylogenetic tree-associated data is often stored in a separate file and needs the expertise to map the data to the tree structure. Lacking standardization to store and represent phylogeny and associated data makes it difficult for researchers to access and integrate the phylogenetic data into their studies. The **treeio** package provides functions to import phylogeny with associated data from several sources, including analysis findings from commonly used software and external data such as experimental data, clinical data, or meta-data. These trees and their associated data can be exported into a single file as `BEAST` or `jtree` formats, and the output file can be parsed back to R by **treeio** and the data is easy to access. The input and output utilities supplied by **treeio** package lay the foundation for phylogenetic data integration for downstream comparative study and visualization. It creates the possibility of integrating a tree with associated data from different sources and extends the applications of phylogenetic analysis in different disciplines.

Part II: Tree data visualization and annotation

Chapter 4

Phylogenetic Tree Visualization

4.1 Introduction

There are many software packages and web tools that are designed for displaying phylogenetic trees, such as **TreeView** (Page, 2002), **FigTree**, **TreeDyn** (Chevenet et al., 2006), *Dendroscope* (Huson & Scornavacca, 2012), **EvolView** (He et al., 2016), and **iTOL** (Letunic & Bork, 2007), *etc.* Only a few of them, such as **FigTree**, **TreeDyn** and **iTOL**, allow users to annotate the trees with colored branches, highlighted clades with tree features. However, their pre-defined annotating functions are usually limited to some specific phylogenetic data. As phylogenetic trees are becoming more widely used in multidisciplinary studies, there is an increasing need to incorporate various types of phylogenetic covariates and other associated data from different sources into the trees for visualizations and further analyses. For instance, the influenza virus has a wide host range, diverse and dynamic genotypes, and characteristic transmission behaviors that are mostly associated with the virus's evolution and essentially among themselves. Therefore, in addition to standalone applications that focus on each of the specific analysis and data types, researchers studying molecular evolution need a robust and programmable platform that allows the high levels of integration and visualization of many of these different aspects of data (raw or from other primary analyses) over the phylogenetic trees to identify their associations and patterns.

To fill this gap, we developed **ggtree** (Yu et al., 2017), a package for the R programming language (R Core Team, 2016) released under the Bioconductor project (Gentleman et al., 2004). The **ggtree** is built to work with `treedata` objects (see Chapters 1 and 9), and display tree graphics with the **ggplot2** package (Wickham, 2016) that was based on the grammar of graphics (Wilkinson et al., 2005).

The R language is increasingly used in phylogenetics. However, a comprehensive package, designed for viewing and annotating phylogenetic trees, particularly with complex data integration, is not yet available. Most of the R packages in phylogenetics focus on specific statistical analyses rather than viewing and annotating the trees with more generalized phylogeny-associated data. Some packages, including **ape** (Paradis et al., 2004) and **phytools** (Revell, 2012), which are capable of displaying and annotating trees, are developed using the base graphics system of R. In particular, **ape** is one of the fundamental packages for phylogenetic analysis and data processing. However, the base graphics system is relatively difficult to extend and limits the complexity of the tree figure to be displayed. **OutbreakTools** (Jombart et al., 2014) and **phyloseq** (McMurdie & Holmes, 2013) extended **ggplot2** to plot phylogenetic trees. The **ggplot2** system of graphics allows rapid customization and exploration of design solutions. However, these packages were designed for epidemiology and microbiome data respectively and did not aim to provide a general solution for tree visualization and annotation. The **ggtree** package also inherits versatile properties of **ggplot2**, and more importantly allows constructing complex tree figures by freely combining multiple layers of annotations (see also Chapter 5) using the tree associated data imported from different sources (see detailed in Chapter 1 and (Wang et al., 2020)).

4.2 Visualizing Phylogenetic Tree with ggtree

The **ggtree** package (Yu et al., 2017) is designed for annotating phylogenetic trees with their associated data of different types and from various sources. These data could come from users or analysis programs and might include evolutionary rates, ancestral sequences, *etc.* that are associated with the taxa from real samples, or with the internal nodes representing hypothetic ancestor strain/species, or with the tree branches indicating evolutionary time courses (Wang et al., 2020). For instance, the data could be the geographic positions of the sampled avian influenza viruses (informed by the survey locations) and the ancestral nodes (by phylogeographic inference) in the viral gene tree (Lam et al., 2012).

The **ggtree** supports **ggplot2**'s graphical language, which allows a high level of customization, is intuitive and flexible. Notably, **ggplot2** itself does not provide low-level geometric objects or other support for tree-like structures, and hence **ggtree** is a useful extension in that regard. Even though the other two phylogenetics-related R packages, **OutbreakTools**, and **phyloseq**, are developed based on **ggplot2**, the most valuable part of the **ggplot2** syntax - adding layers of annotations - is not supported in these packages. For example, if we have plotted a tree without taxa labels, **OutbreakTools** and **phyloseq** provide no easy way for general `R` users, who have little knowledge about the infrastructures of these packages, to add a layer of taxa labels. The **ggtree** extends **ggplot2** to support tree objects by implementing a geometric layer, `geom_tree()`, to support visualizing tree structure. In **ggtree**, viewing a phylogenetic tree is relatively easy, via the command `ggplot(tree_object) + geom_tree() + theme_tree()` or `ggtree(tree_object)` for short. Layers of annotations can be added one-by-one via the `+` operator. To facilitate tree visualization, **ggtree** provides several geometric layers, including `geom_treescale()` for adding legend of tree branch scale (genetic distance, divergence time, *etc.*), `geom_range()` for displaying uncertainty of branch lengths (confidence interval or range, *etc.*), `geom_tiplab()` for adding taxa label, `geom_tippoint()` and `geom_nodepoint()` for adding symbols of tips and internal nodes, `geom_hilight()` for highlighting clades with rectangle, and `geom_cladelab()` for annotating selected clades with bar and text label, *etc.*. A full list of geometric layers provided by **ggtree** is summarized in Table 5.1.

To view a phylogenetic tree, we first need to parse the tree file into `R`. The **treeio** package is able to parse diverse annotation data from different software outputs into `S4` phylogenetic data objects (see also Chapter 1). The **ggtree** package mainly utilizes these `S4` objects to display and annotate the tree. Other R packages defined `S3`/`S4` classes to store phylogenetic trees with domain-specific associated data, including `phylo4` and `phylo4d` defined in the **phylobase** package, `obkdata` defined in the **OutbreakTools** package, and `phyloseq` defined in the **phyloseq** package, *etc.* All these tree objects are also supported in **ggtree** and their specific annotation data can be used to annotate the tree directly in **ggtree** (see also Chapter 9). Such compatibility of **ggtree** facilitates the integration of data and analysis results. In addition, **ggtree** also supports other tree-like structures, including dendrogram and tree graphs.

4.2.1 Basic Tree Visualization

The **ggtree** package extends **ggplot2** (Wickham, 2016) package to support viewing a phylogenetic tree. It implements `geom_tree()` layer for displaying a phylogenetic tree, as shown below in Figure 4.1A.

```
library("treeio")
library("ggtree")

nwk <- system.file("extdata", "sample.nwk", package="treeio")
tree <- read.tree(nwk)

ggplot(tree, aes(x, y)) + geom_tree() + theme_tree()
```

The function, `ggtree()`, was implemented as a shortcut to visualize a tree, and it works exactly the same as shown above.

The **ggtree** package takes all the advantages of **ggplot2**. For example, we can change the color, size, and type of the lines as we do with **ggplot2** (Figure 4.1B).

```
ggtree(tree, color="firebrick", size=2, linetype="dotted")
```

By default, the tree is viewed in ladderize form, user can set the parameter `ladderize = FALSE` to disable it (Figure 4.1C, see also FAQ A.5).

```
ggtree(tree, ladderize=FALSE)
```

The `branch.length` is used to scale the edge, user can set the parameter `branch.length = "none"` to only view the tree topology (cladogram, Figure 4.1D)

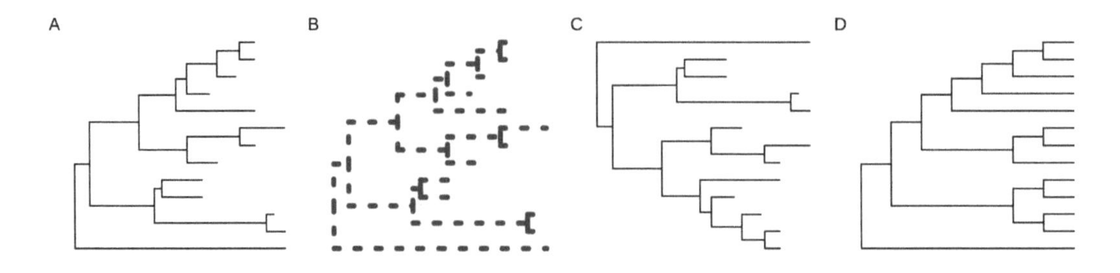

Figure 4.1: **Basic tree visualization.** Default ggtree output with ladderized effect (A), non-variable setting (*e.g.*, color, size, line type) (B), non-ladderized tree (C), cladogram that only displays tree topology without branch length information (D).

or other numerical variables to scale the tree (*e.g.*, d_N/d_S, see also in Chapter 5).

```
ggtree(tree, branch.length="none")
```

4.2.2 Layouts of a phylogenetic tree

Viewing phylogenetic with **ggtree** is quite simple, just pass the tree object to the `ggtree()` function. We have developed several types of layouts for tree presentation (Figure 4.2), including *rectangular* (by default), *roundrect* (rounded rectangular), *ellipse*, *slanted*, *circular*, *fan*, *unrooted* (equal angle and daylight methods), time-scaled, and two-dimensional layouts.

Here are examples of visualizing a tree with different layouts:

```
library(ggtree)
set.seed(2017-02-16)
tree <- rtree(50)
ggtree(tree)
ggtree(tree, layout="roundrect")
ggtree(tree, layout="slanted")
ggtree(tree, layout="ellipse")
ggtree(tree, layout="circular")
ggtree(tree, layout="fan", open.angle=120)
ggtree(tree, layout="equal_angle")
ggtree(tree, layout="daylight")
ggtree(tree, branch.length='none')
ggtree(tree, layout="ellipse", branch.length="none")
ggtree(tree, branch.length='none', layout='circular')
ggtree(tree, layout="daylight", branch.length = 'none')
```

Other possible layouts that can be drawn by modifying scales/coordination (Figure 4.3).

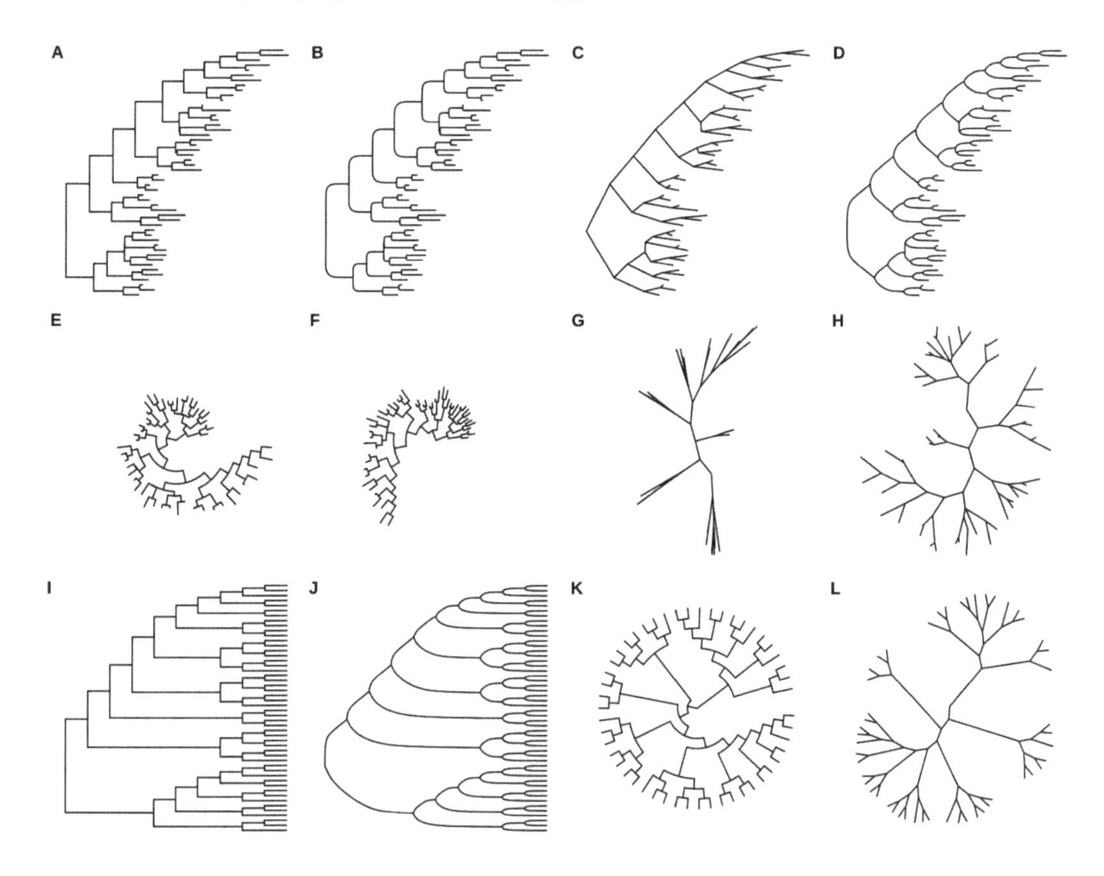

Figure 4.2: **Tree layouts.** Phylogram: rectangular layout (A), rounded rectangular layout (B), slanted layout (C), ellipse layout (D), circular layout (E), and fan layout (F). Unrooted: equal-angle method (G) and daylight method (H). Cladogram: rectangular layout (I), ellipse (J), circular layout (K), and unrooted layout (L). Slanted and fan layouts for cladogram are also supported.

```
ggtree(tree) + scale_x_reverse()
ggtree(tree) + coord_flip()
ggtree(tree) + layout_dendrogram()
ggplotify::as.ggplot(ggtree(tree), angle=-30, scale=.9)
ggtree(tree, layout='slanted') + coord_flip()
ggtree(tree, layout='slanted', branch.length='none') +
layout_dendrogram()
ggtree(tree, layout='circular') + xlim(-10, NA)
ggtree(tree) + layout_inward_circular()
ggtree(tree) + layout_inward_circular(xlim=15)
```

Phylogram. Layouts of *rectangular*, *roundrect*, *slanted*, *ellipse*, *circular*, and *fan* are supported to visualize phylogram (by default, with branch length scaled) as demonstrated in Figures 4.2A-F.

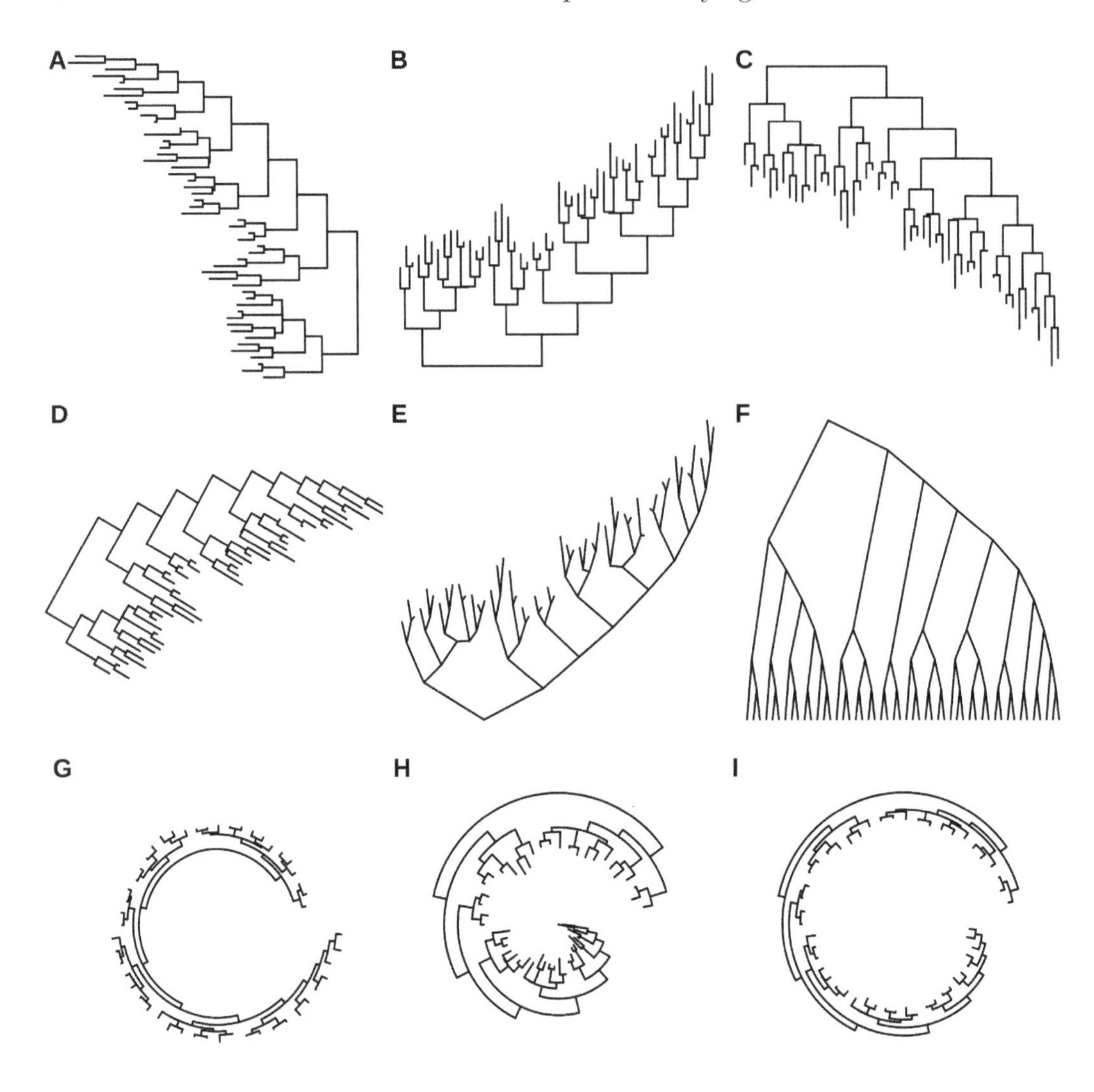

Figure 4.3: **Derived Tree layouts.** right-to-left rectangular layout (A), bottom-up rectangular layout (B), top-down rectangular layout (Dendrogram) (C), rotated rectangular layout (D), bottom-up slanted layout (E), top-down slanted layout (Cladogram) (F), circular layout (G), circular inward layout (H and I).

Unrooted layout. Unrooted (also called `radial') layout is supported by equal-angle and daylight algorithms; users can specify unrooted layout algorithm by passing "equal_angle" or "daylight" tolayout‘` parameter to visualize the tree. The equal-angle method was proposed by Christopher Meacham in *PLOTREE*, which was incorporated in **PHYLIP** (Retief, 2000). This method starts from the root of the tree and allocates arcs of angle to each subtree proportional to the number of tips in it. It iterates from root to tips and subdivides the angle allocated to a subtree into angles for its dependent subtrees. This method is fast and was implemented in many software packages. As shown in Figure 4.2G, the equal angle method has a drawback that tips tend to be clustered together, which will leave many spaces unused. The daylight method starts from an initial tree built by equal angle and iteratively improves it by successively going to each interior node and swinging subtrees so that the arcs of "daylight" are equal (Figure 4.2H). This method was firstly implemented in **PAUP*** (Wilgenbusch & Swofford, 2003).

Cladogram. To visualize a cladogram that is without branch length scaling and only displays the tree structure, `branch.length` is set to "none" and it works for all types of layouts (Figures 4.2I-L).

Timescaled layout. For a timescaled tree, the most recent sampling date must be specified via the `mrsd` parameter, and `ggtree()` will scale the tree by sampling (tip) and divergence (internal node) time, and a timescale axis will be displayed under the tree by default. Users can use the **deeptime** package to add geologic timescale (e.g., periods and eras) to a `ggtree()` plot.

```
beast_file <- system.file("examples/MCC_FluA_H3.tree",
                          package="ggtree")
beast_tree <- read.beast(beast_file)
ggtree(beast_tree, mrsd="2013-01-01") + theme_tree2()
```

Two-dimensional tree layout. A two-dimensional tree is a projection of the phylogenetic tree in a space defined by the associated phenotype (numerical or categorical trait, on the y-axis) and tree branch scale (*e.g.*, evolutionary distance, divergent time, on the x-axis). The phenotype can be a measure of certain biological characteristics of the taxa and hypothetical ancestors in the tree. This layout is useful to track the virus phenotypes or other behaviors (y-axis) changing with the virus evolution (x-axis). In fact, the analysis of phenotypes or genotypes over evolutionary time have been widely used for study of influenza virus evolution (Neher et al., 2016), though such analysis diagrams are not tree-like, *i.e.*, no connection between data points, unlike our two-dimensional tree layout that connects data points with the corresponding tree branches. Therefore, this new layout we provided will make such data analysis easier and more scalable for large sequence datasets.

In this example, we used the previous timescaled tree of H3 human and swine influenza viruses (Figure 4.4; data published in (Liang et al., 2014)) and scaled the y-axis based on the predicted N-linked glycosylation sites (NLG) for each of the taxon and ancestral sequences of hemagglutinin proteins. The NLG sites were predicted using the NetNGlyc 1.0 Server. To scale the y-axis, the parameter `yscale` in the `ggtree()` function is set to a numerical or categorical variable. If `yscale` is a categorical variable as in this example, users should specify how the categories are to be mapped to numerical values via the `yscale_mapping` variables.

```
NAG_file <- system.file("examples/NAG_inHA1.txt", package="ggtree")

NAG.df <- read.table(NAG_file, sep="\t", header=FALSE,
                     stringsAsFactors = FALSE)
```

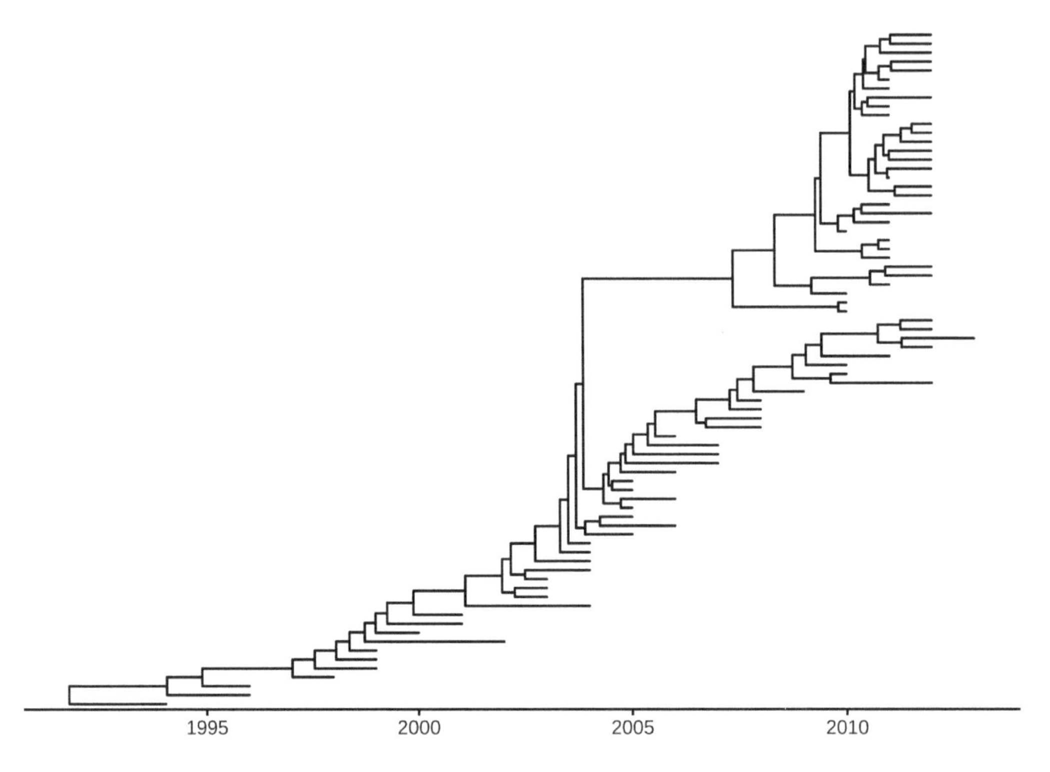

Figure 4.4: **Timescaled layout.** The x-axis is the timescale (in units of the year). The divergence time in this example was inferred by **BEAST** using the molecular clock model.

```
NAG <- NAG.df[,2]
names(NAG) <- NAG.df[,1]

## separate the tree by host species
tip <- as.phylo(beast_tree)$tip.label
beast_tree <- groupOTU(beast_tree, tip[grep("Swine", tip)],
                       group_name = "host")

p <- ggtree(beast_tree, aes(color=host), mrsd="2013-01-01",
            yscale = "label", yscale_mapping = NAG) +
  theme_classic() + theme(legend.position='none') +
  scale_color_manual(values=c("blue", "red"),
                     labels=c("human", "swine")) +
  ylab("Number of predicted N-linked glycosylation sites")

## (optional) add more annotations to help interpretation
p + geom_nodepoint(color="grey", size=3, alpha=.8) +
  geom_rootpoint(color="black", size=3) +
  geom_tippoint(size=3, alpha=.5) +
```

```
annotate("point", 1992, 5.6, size=3, color="black") +
annotate("point", 1992, 5.4, size=3, color="grey") +
annotate("point", 1991.6, 5.2, size=3, color="blue") +
annotate("point", 1992, 5.2, size=3, color="red") +
annotate("text", 1992.3, 5.6, hjust=0, size=4, label="Root node") +
annotate("text", 1992.3, 5.4, hjust=0, size=4,
        label="Internal nodes") +
annotate("text", 1992.3, 5.2, hjust=0, size=4,
        label="Tip nodes (blue: human; red: swine)")
```

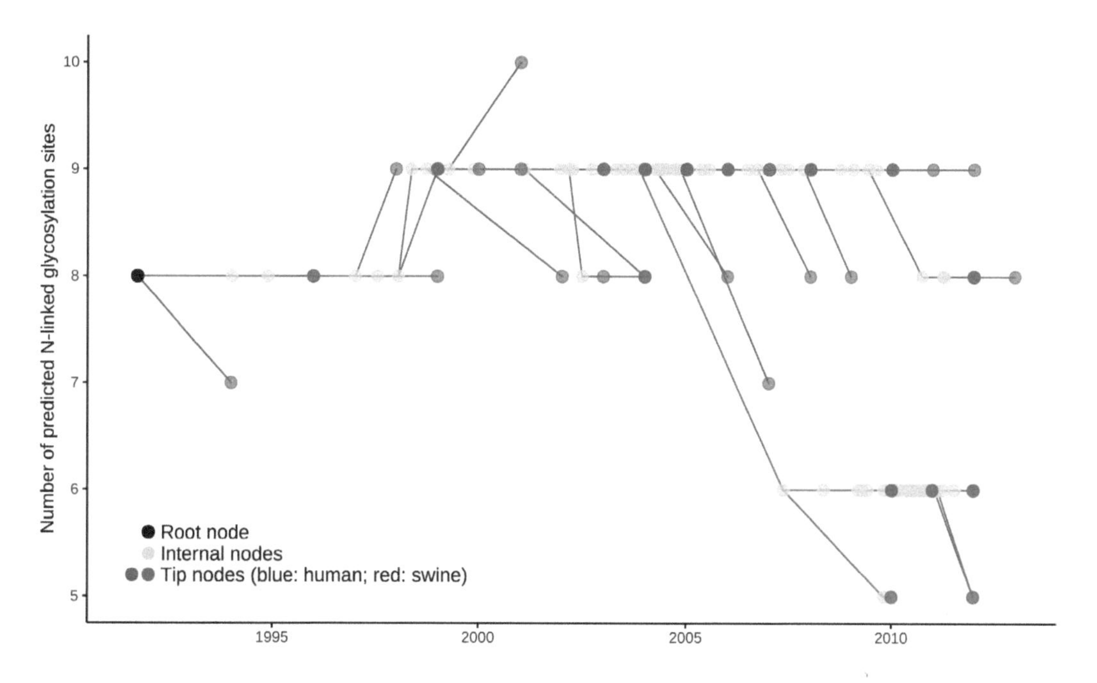

Figure 4.5: **Two-dimensional tree layout.** The trunk and other branches are highlighted in red (for swine) and blue (for humans). The x-axis is scaled to the branch length (in units of year) of the timescaled tree. The y-axis is scaled to the node attribute variable, in this case, the number of predicted N-linked glycosylation sites (NLG) on the hemagglutinin protein. Colored circles indicate the different types of tree nodes. Note that nodes assigned the same x- (temporal) and y- (NLG) coordinates are superimposed in this representation and appear as one node, which is shaded based on the colors of all the nodes at that point.

As shown in Figure 4.5, a two-dimensional tree is good at visualizing the change of phenotype over the evolution in the phylogenetic tree. In this example, it is shown that the H3 gene of the human influenza A virus maintained a high level of N-linked glycosylation sites (n=8 to 9) over the last two decades and dropped significantly to 5 or 6 in a separate viral lineage transmitted to swine populations and established there. It was indeed hypothesized that the human influenza virus with a high level of glycosylation on the viral hemagglutinin protein provides a better shielding effect to protect the antigenic sites from exposure to the herd immunity, and thus has a selective advantage in human populations that maintain a high level of herd immunity against the circulating human influenza virus strains. For the viral lineage that newly jumped across the species barrier and transmitted to the swine population, the shielding effect of the high-level surface glycan oppositely imposes selective disadvantage because the receptor-binding domain may also be shielded which greatly affects the viral fitness of the lineage that newly adapted to a new host species. Another example of a two-dimensional tree can be found in Figure 4.12.

4.3 Displaying Tree Components

4.3.1 Displaying treescale (evolution distance)

To show treescale, the user can use `geom_treescale()` layer (Figures 4.6A-C).

```
ggtree(tree) + geom_treescale()
```

`geom_treescale()` supports the following parameters:

- x and y for treescale position
- *width* for the length of the treescale
- *fontsize* for the size of the text
- *linesize* for the size of the line
- *offset* for relative position of the line and the text
- *color* for color of the treescale

```
ggtree(tree) + geom_treescale(x=0, y=45, width=1, color='red')
ggtree(tree) + geom_treescale(fontsize=6, linesize=2, offset=1)
```

We can also use `theme_tree2()` to display the treescale by adding x *axis* (Figure 4.6D).

```
ggtree(tree) + theme_tree2()
```

Treescale is not restricted to evolution distance, **treeio** can rescale the tree with other numerical variables (details described in session 2.4), and **ggtree** allows users to specify a numerical variable to serve as branch length for visualization (details described in session 4.3).

4.3.2 Displaying nodes/tips

Showing all the internal nodes and tips in the tree can be done by adding a layer of points using `geom_nodepoint()`, `geom_tippoint()`, or `geom_point()` (Figure 4.7).

```
ggtree(tree) +
    geom_point(aes(shape=isTip, color=isTip), size=3)

p <- ggtree(tree) +
```

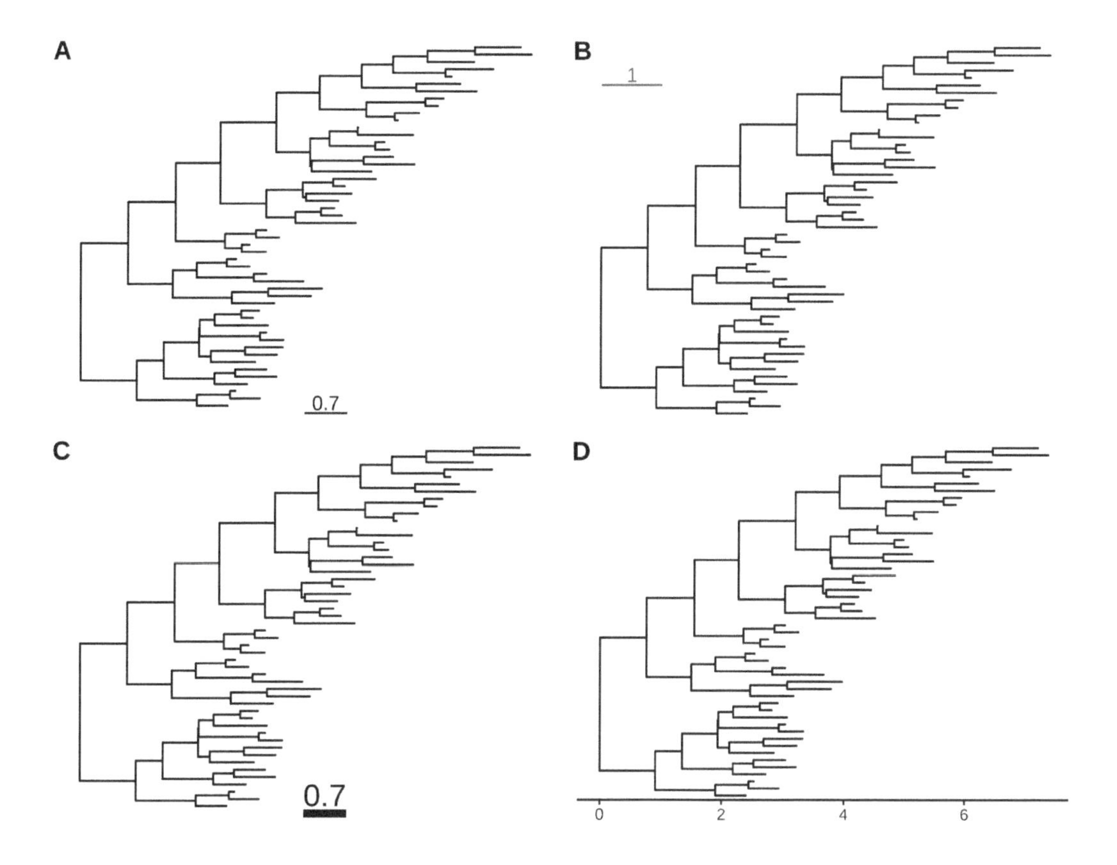

Figure 4.6: **Display treescale.** `geom_treescale()` automatically add a scale bar for evolutionary distance (A). Users can modify the color, width, and position of the scale (B) as well as the size of the scale bar and text and their relative position (C). Another possible solution is to enable the x-axis which is useful for the timescaled tree (D).

```
    geom_nodepoint(color="#b5e521", alpha=1/4, size=10)
p + geom_tippoint(color="#FDAC4F", shape=8, size=3)
```

4.3.3 Displaying labels

Users can use `geom_text()` or `geom_label()` to display the node (if available) and tip labels simultaneously or `geom_tiplab()` to only display tip labels (Figure 4.8A).

```
p + geom_tiplab(size=3, color="purple")
```

The `geom_tiplab()` layer not only supports using *text* or *label* geom to display labels, but it also supports *image* geom to label tip with image files (see Chapter 7). A corresponding geom, `geom_nodelab()` is also provided for displaying node labels.

For *circular* and *unrooted* layouts, **ggtree** supports rotating node labels according to the angles of the branches (Figure 4.8B).

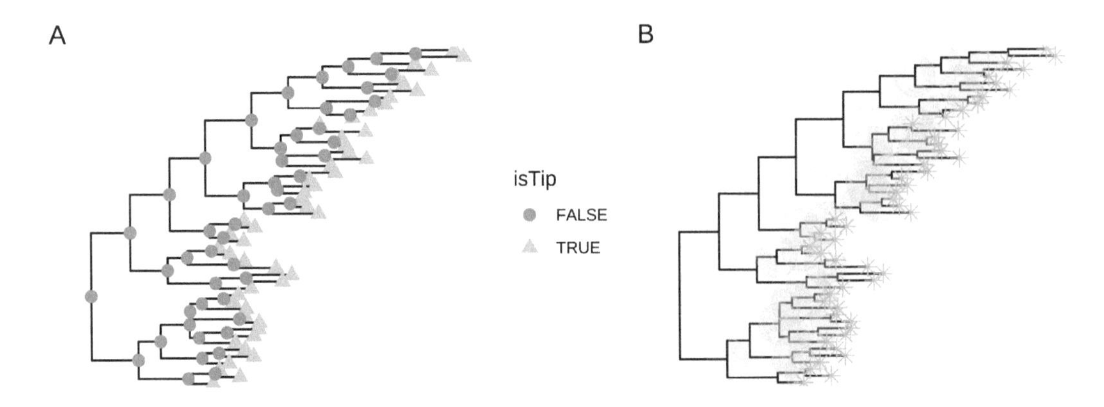

Figure 4.7: **Display external and internal nodes.** `geom_point()` automatically add symbolic points of all nodes (A). `geom_nodepoint()` adds symbolic points for internal nodes and `geom_tippoint()` adds symbolic points for external nodes (B).

```
ggtree(tree, layout="circular") + geom_tiplab(aes(angle=angle),
color='blue')
```

For long tip labels, the label may be truncated. There are several ways to solve this issue (see FAQ: Tip label truncated). Another solution to solve this issue is to display tip labels as y-axis labels (Figure 4.8C). However, it only works for rectangular and dendrogram layouts and users need to use `theme()` to adjust tip labels in this case.

```
ggtree(tree) + geom_tiplab(as_ylab=TRUE, color='firebrick')
```

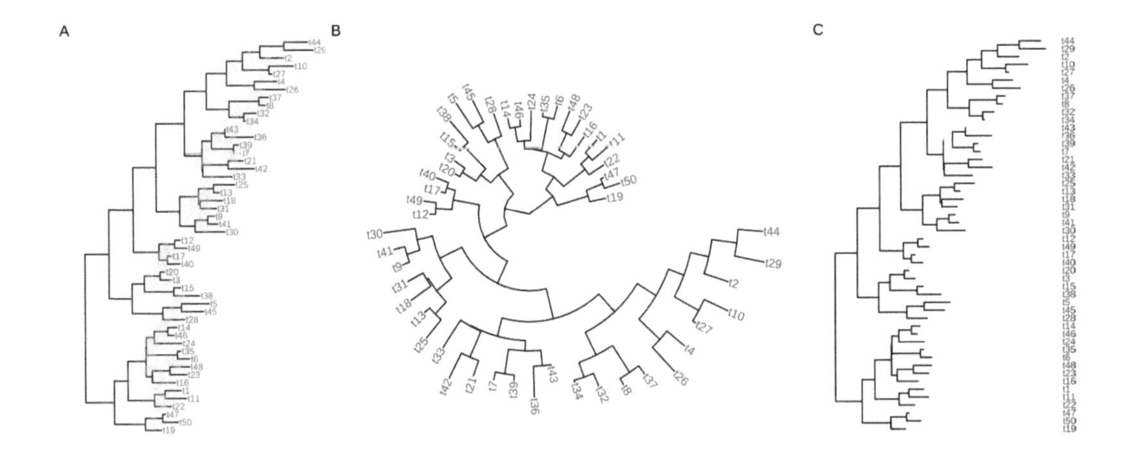

Figure 4.8: **Display tip labels.** `geom_tiplab()` supports displaying tip labels (A). For the circular, fan, or unrooted tree layouts, the labels can be rotated to fit the angle of the branches (B). For dendrogram/rectangular layout, tip labels can be displayed as y-axis labels (C).

By default, the positions to display text are based on the node positions; we can change them to be based on the middle of the branch/edge (by setting `aes(x = branch)`), which is very useful when annotating transition from the parent node to the child node.

4.3.4 Displaying root-edge

The `ggtree()` doesn't plot the root-edge by default. Users can use `geom_rootedge()` to automatically display the root-edge (Figure 4.9A). If there is no root edge information, `geom_rootedge()` will display nothing by default (Figure 4.9B). Users can set the root-edge to the tree (Figure 4.9C) or specify `rootedge` in `geom_rootedge()` (Figure 4.9D). A long root length is useful to increase readability of the circular tree (see also FAQ: Enlarge center space).

```
## with root-edge = 1
tree1 <- read.tree(text='((A:1,B:2):3,C:2):1;')
ggtree(tree1) + geom_tiplab() + geom_rootedge()

## without root-edge
tree2 <- read.tree(text='((A:1,B:2):3,C:2);')
ggtree(tree2) + geom_tiplab() + geom_rootedge()

## setting root-edge
tree2$root.edge <- 2
ggtree(tree2) + geom_tiplab() + geom_rootedge()

## specify the length of root edge for just plotting
## this will ignore tree$root.edge
ggtree(tree2) + geom_tiplab() + geom_rootedge(rootedge = 3)
```

4.3.5 Color tree

In **ggtree** (Yu et al., 2018), coloring phylogenetic tree is easy, by using `aes(color=VAR)` to map the color of the tree based on a specific variable (both numerical and categorical variables are supported, see Figure 4.10).

```
ggtree(beast_tree, aes(color=rate)) +
    scale_color_continuous(low='darkgreen', high='red') +
    theme(legend.position="right")
```

Users can use any feature (if available), including clade posterior and d_N/d_S, *etc.*, to scale the color of the tree. If the feature is a continuous numerical value, **ggtree** provides a `continuous` parameter to support plotting continuous state transition in edges. Here, we use an example[1] to demonstrate this functionality (Figure 4.11A). If you want to add a thin black border in tree branches, you can place a tree with

[1]http://www.phytools.org/eqg2015/asr.html

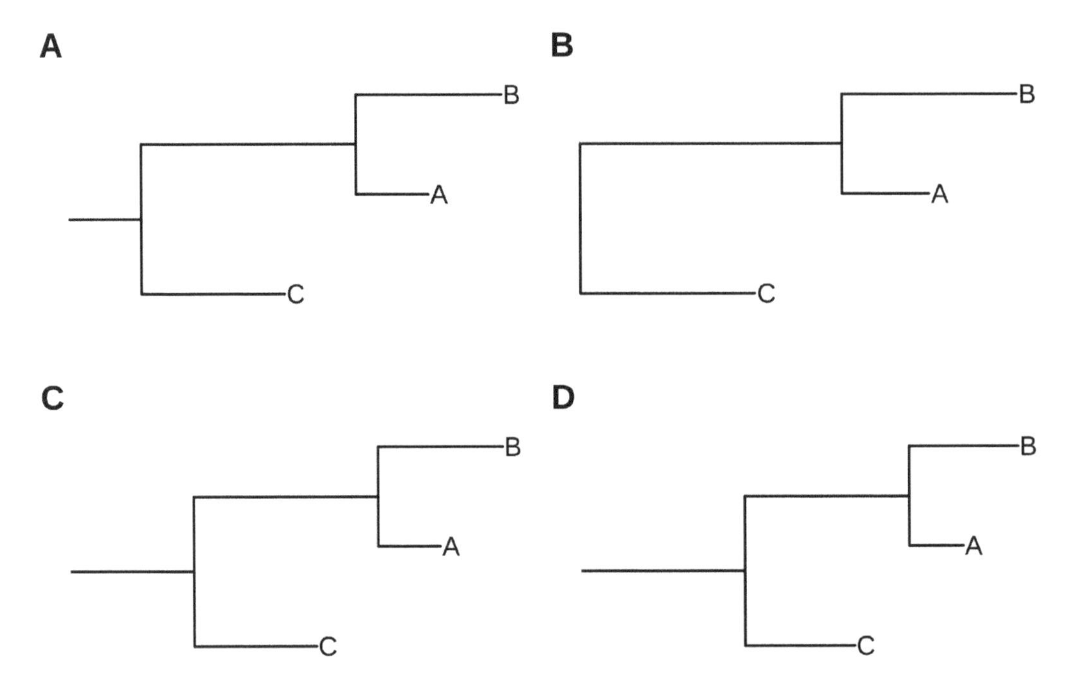

Figure 4.9: **Display root-edge.** `geom_rootedge()` supports displaying root-edge if the root edge was presented (A). It shows nothing if there is no root-edge (B). In this case, users can manually set the root edge for the tree (C) or just specify the length of the root for plotting (D).

black and slightly thicker branches below your tree to emulate edge outlines as demonstrated in Figure 4.11B.

```
library(ggtree)
library(treeio)
library(tidytree)
library(ggplot2)
library(TDbook)
## ref: http://www.phytools.org/eqg2015/asr.html
##
## load `tree_anole` and `df_svl` from 'TDbook'
svl <- as.matrix(df_svl)[,1]
fit <- phytools::fastAnc(tree_anole, svl, vars=TRUE, CI=TRUE)

td <- data.frame(node = nodeid(tree_anole, names(svl)),
               trait = svl)
nd <- data.frame(node = names(fit$ace), trait = fit$ace)

d <- rbind(td, nd)
d$node <- as.numeric(d$node)
tree <- full_join(tree_anole, d, by = 'node')
```

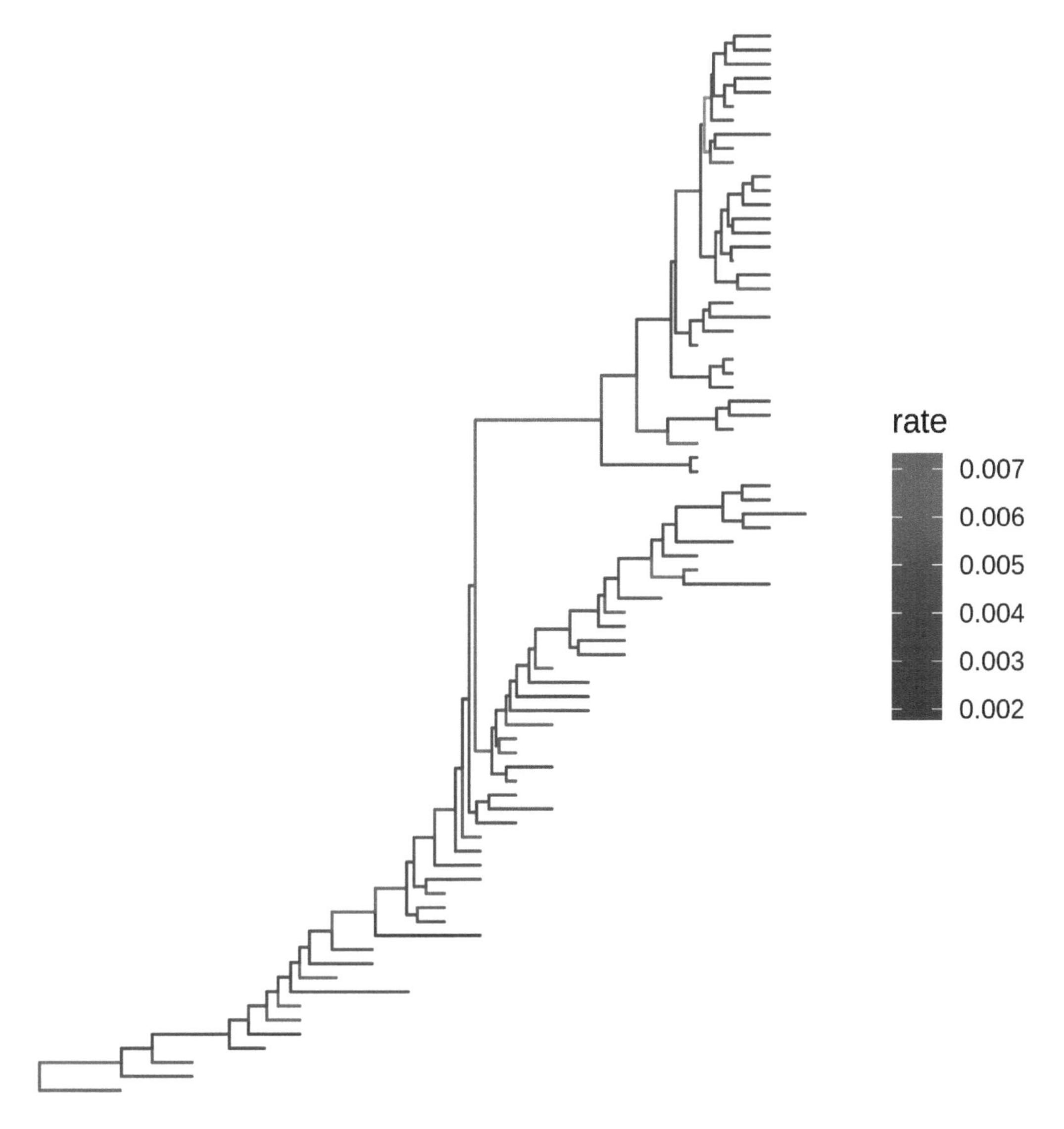

Figure 4.10: **Color tree by continuous or discrete feature.** Edges are colored by values associated with the child nodes.

```
p1 <- ggtree(tree, aes(color=trait), layout = 'circular',
        ladderize = FALSE, continuous = 'colour', size=2) +
    scale_color_gradientn(colours=c("red", 'orange', 'green', 'cyan',
    'blue')) +
    geom_tiplab(hjust = -.1) +
    xlim(0, 1.2) +
    theme(legend.position = c(.05, .85))

p2 <- ggtree(tree, layout='circular', ladderize = FALSE, size=2.8) +
    geom_tree(aes(color=trait), continuous = 'colour', size=2) +
```

```
    scale_color_gradientn(colours=c("red", 'orange', 'green', 'cyan',
    'blue')) +
    geom_tiplab(aes(color=trait), hjust = -.1) +
    xlim(0, 1.2) +
    theme(legend.position = c(.05, .85))

plot_list(p1, p2, ncol=2, tag_levels="A")
```

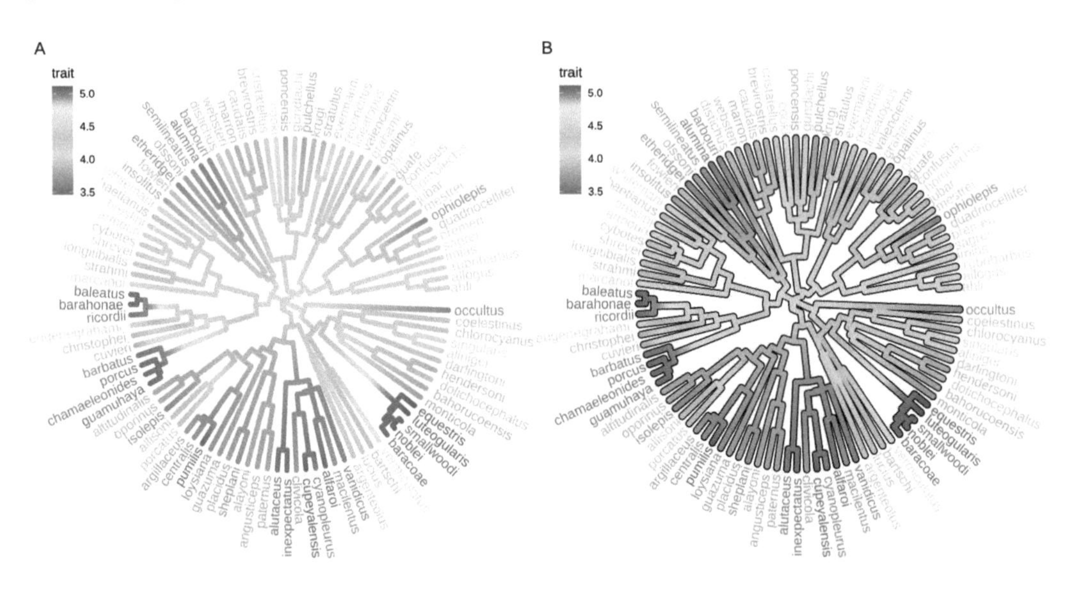

Figure 4.11: **Continuous state transition in edges.** Edges are colored by the values from ancestral trait to offspring.

Besides, we can use a two-dimensional tree (as demonstrated in Figure 4.5) to visualize phenotype on the vertical dimension to create the phenogram Figure 4.12. We can use the **ggrepel** package to repel tip labels to avoid overlapping as demonstrated in Figure A.4.

```
ggtree(tree, aes(color=trait), continuous = 'colour', yscale = "trait") +
    scale_color_viridis_c() + theme_minimal()
```

4.3.6 Rescale tree

Most of the phylogenetic trees are scaled by evolutionary distance (substitution/site). In **ggtree**, users can rescale a phylogenetic tree by any numerical variable inferred by evolutionary analysis (*e.g.*, d_N/d_S).

This example displays a time tree (Figure 4.13A) and the branches were rescaled by substitution rate inferred by BEAST (Figure 4.13B).

```
library("treeio")
beast_file <- system.file("examples/MCC_FluA_H3.tree", package="ggtree")
```

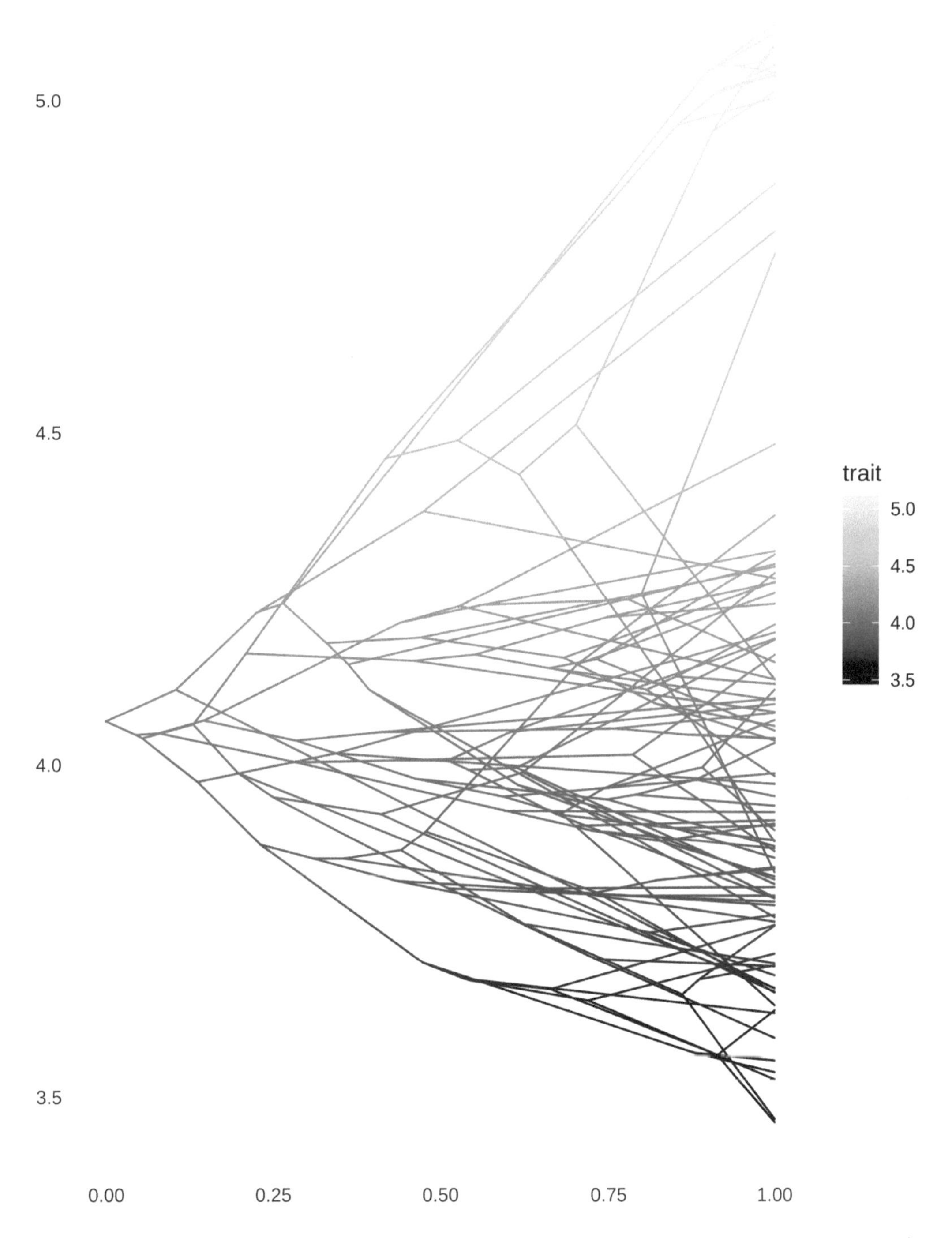

Figure 4.12: **Phenogram.** Projecting the tree into a space defined by time (or genetic distance) on the horizontal axis and phenotype on the vertical dimension.

```
beast_tree <- read.beast(beast_file)
beast_tree

## 'treedata' S4 object that stored information
## of
##  '/home/ygc/R/library/ggtree/examples/MCC_FluA_H3.tree'.
##
## ...@ phylo:
##
## Phylogenetic tree with 76 tips and 75 internal nodes.
##
## Tip labels:
##   A/Hokkaido/30-1-a/2013, A/New_York/334/2004,
## A/New_York/463/2005, A/New_York/452/1999,
## A/New_York/238/2005, A/New_York/523/1998, ...
##
## Rooted; includes branch lengths.
##
## with the following features available:
##   'height', 'height_0.95_HPD', 'height_median',
## 'height_range', 'length', 'length_0.95_HPD',
## 'length_median', 'length_range', 'posterior', 'rate',
## 'rate_0.95_HPD', 'rate_median', 'rate_range'.
```

```
p1 <- ggtree(beast_tree, mrsd='2013-01-01') + theme_tree2() +
    labs(caption="Divergence time")
p2 <- ggtree(beast_tree, branch.length='rate') + theme_tree2() +
    labs(caption="Substitution rate")
```

The following example draws a tree inferred by CodeML (Figure 4.13C), and the branches can be rescaled by using d_N/d_S values (Figure 4.13D).

```
mlcfile <- system.file("extdata/PAML_Codeml", "mlc", package="treeio")
mlc_tree <- read.codeml_mlc(mlcfile)
p3 <- ggtree(mlc_tree) + theme_tree2() +
    labs(caption="nucleotide substitutions per codon")
p4 <- ggtree(mlc_tree, branch.length='dN_vs_dS') + theme_tree2() +
    labs(caption="dN/dS tree")
```

This provides a very convenient way to allow us to explore the relationship between tree associated data and tree structure through visualization. In addition to specifying `branch.length` in tree visualization, users can change branch length stored in tree object by using `rescale_tree()` function provided by the **treeio** package (Wang et al., 2020), and the following command will display a tree that is identical to Figure 4.13B. The `rescale_tree()` function was documented in session 2.4.

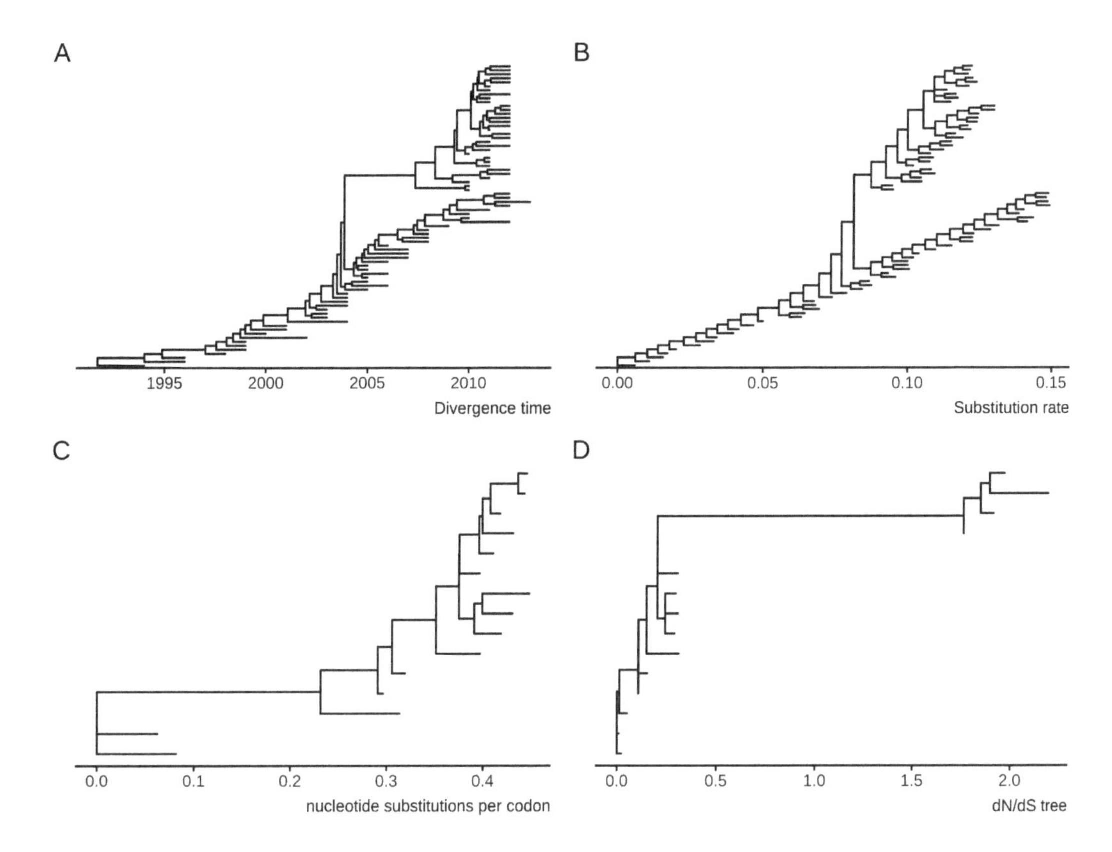

Figure 4.13: **Rescale tree branches.** A time-scaled tree inferred by BEAST (A) and its branches were rescaled by substitution rate (B). A tree was inferred by CodeML (C) and the branches were rescaled by d_N/d_S values (D).

```
beast_tree2 <- rescale_tree(beast_tree, branch.length='rate')
ggtree(beast_tree2) + theme_tree2()
```

4.3.7 Modify components of a theme

The `theme_tree()` defined a totally blank canvas, while `theme_tree2()` adds phylogenetic distance (via x-axis). These two themes all accept a parameter of `bgcolor` that defined the background color. Users can use any theme components to the `theme_tree()` or `theme_tree2()` functions to modify them (Figure 4.14).

```
set.seed(2019)
x <- rtree(30)
ggtree(x, color="#0808E5", size=1) + theme_tree("#FEE4E9")
ggtree(x, color="orange", size=1) + theme_tree('grey30')
```

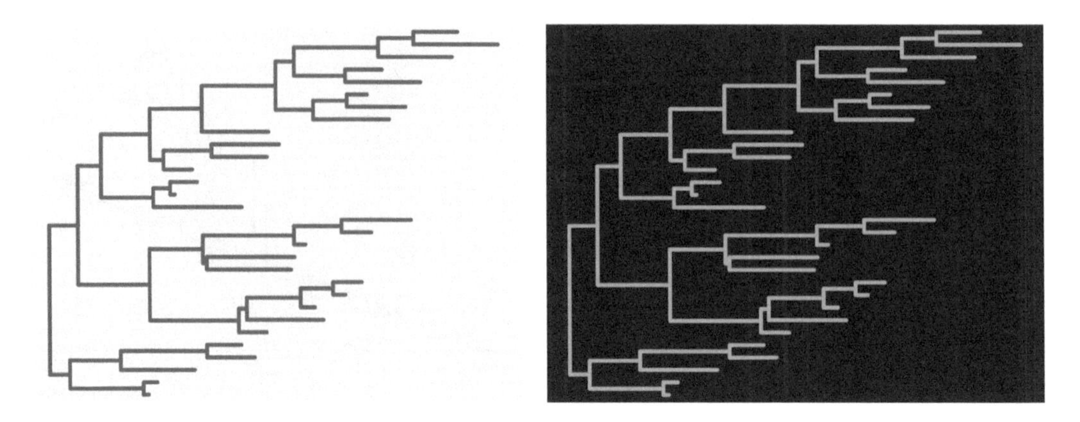

Figure 4.14: **Three themes.** All ggplot2 theme components can be modified, and all the ggplot2 themes can be applied to `ggtree()` output.

Users can also use an image file as the tree background, see example in Appendix B.

4.4 Visualize a List of Trees

The **ggtree** supports `multiPhylo` and `treedataList` objects and a list of trees can be viewed simultaneously. The trees will visualize one on top of another and can be plotted in different panels through `facet_wrap()` or `facet_grid()` functions (Figure 4.15).

```
## trees <- lapply(c(10, 20, 40), rtree)
## class(trees) <- "multiPhylo"
## ggtree(trees) + facet_wrap(~.id, scale="free") + geom_tiplab()

f <- system.file("extdata/r8s", "H3_r8s_output.log", package="treeio")
r8s <- read.r8s(f)
ggtree(r8s) + facet_wrap( ~.id, scale="free") + theme_tree2()
```

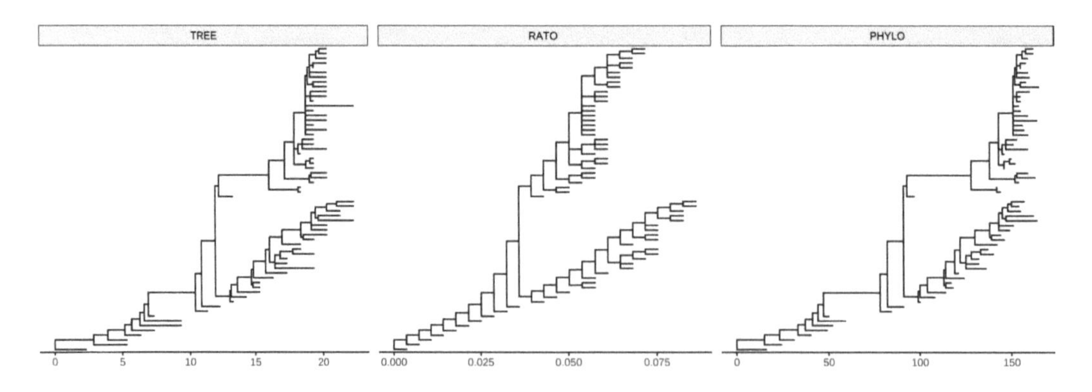

Figure 4.15: **Visualizing multiPhylo object.** The `ggtree()` function supports visualizing multiple trees stored in the `multiPhylo` or `treedataList` objects.

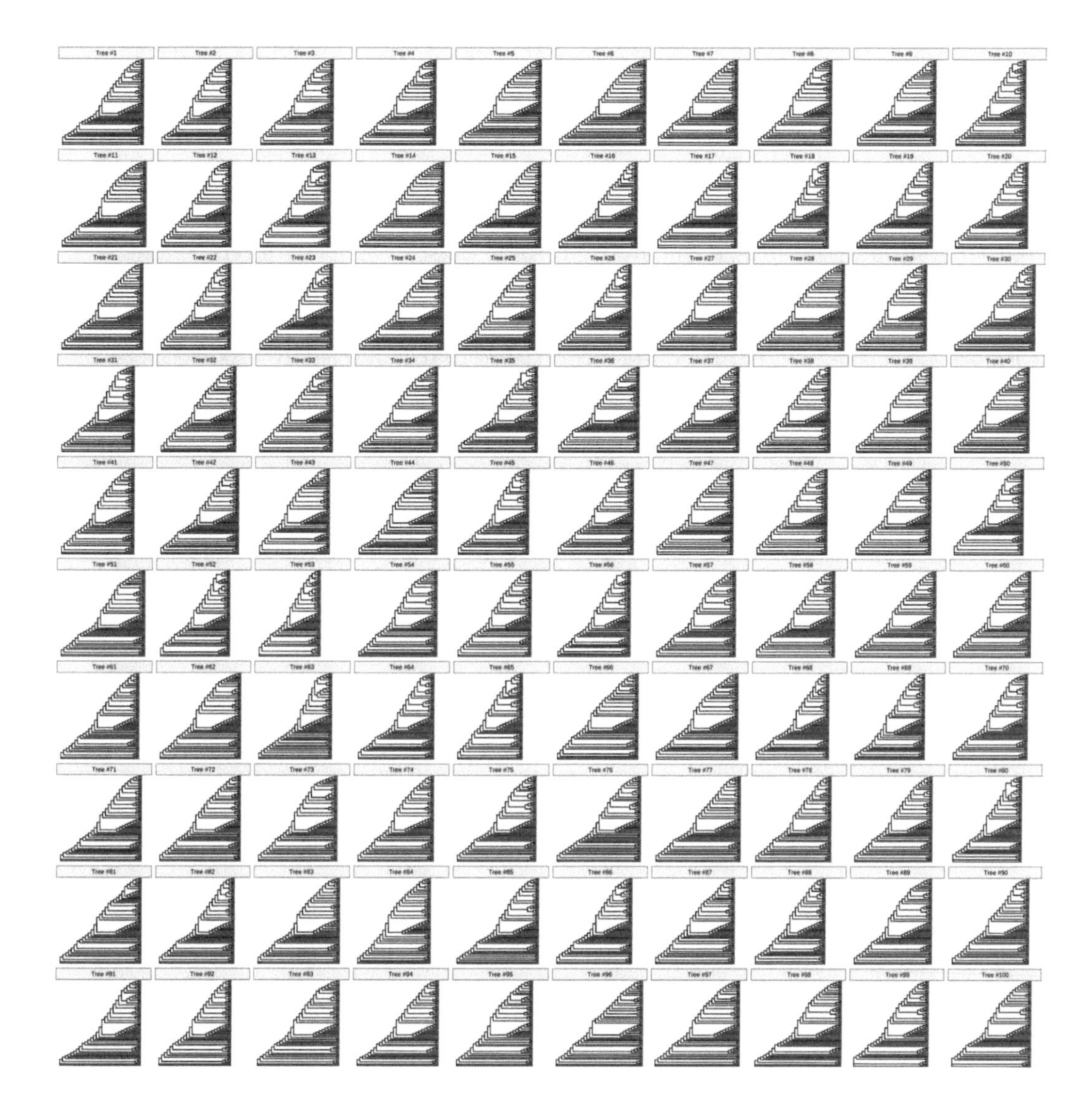

Figure 4.16: **Visualizing one hundred bootstrap trees simultaneously.**

One hundred bootstrap trees can also be viewed simultaneously (Figure 4.16). This allows researchers to explore a large set of phylogenetic trees to find consensus and distinct trees. The consensus tree can be summarized via a density tree (Figure 4.18).

```
btrees <- read.tree(system.file("extdata/RAxML",
                                "RAxML_bootstrap.H3",
                                package="treeio")
                    )
ggtree(btrees) + facet_wrap(~.id, ncol=10)
```

4.4.1 Annotate one tree with values from different variables

To annotate one tree (the same tree) with the values from different variables, one can plot them separately and use **patchwork** or **aplot** to combine them side-by-side.

Another solution is to utilize the ability to plot a list of trees by **ggtree**, and then add annotation layers for the selected variable at a specific panel via the subset aesthetic mapping supported by **ggtree** or using the `td_filter()` as demonstrated in Figure 4.17. The `.id` is the conserved variable that is internally used to store the IDs of different trees.

```
set.seed(2020)
x <- rtree(30)
d <- data.frame(label=x$tip.label, var1=abs(rnorm(30)),
var2=abs(rnorm(30)))
tree <- full_join(x, d, by='label')
trs <- list(TREE1 = tree, TREE2 = tree)
class(trs) <- 'treedataList'
ggtree(trs) + facet_wrap(~.id) +
  geom_tippoint(aes(subset=.id == 'TREE1', colour=var1)) +
  scale_colour_gradient(low='blue', high='red') +
  ggnewscale::new_scale_colour()  +
  geom_tippoint(aes(colour=var2), data=td_filter(.id == "TREE2")) +
  scale_colour_viridis_c()
```

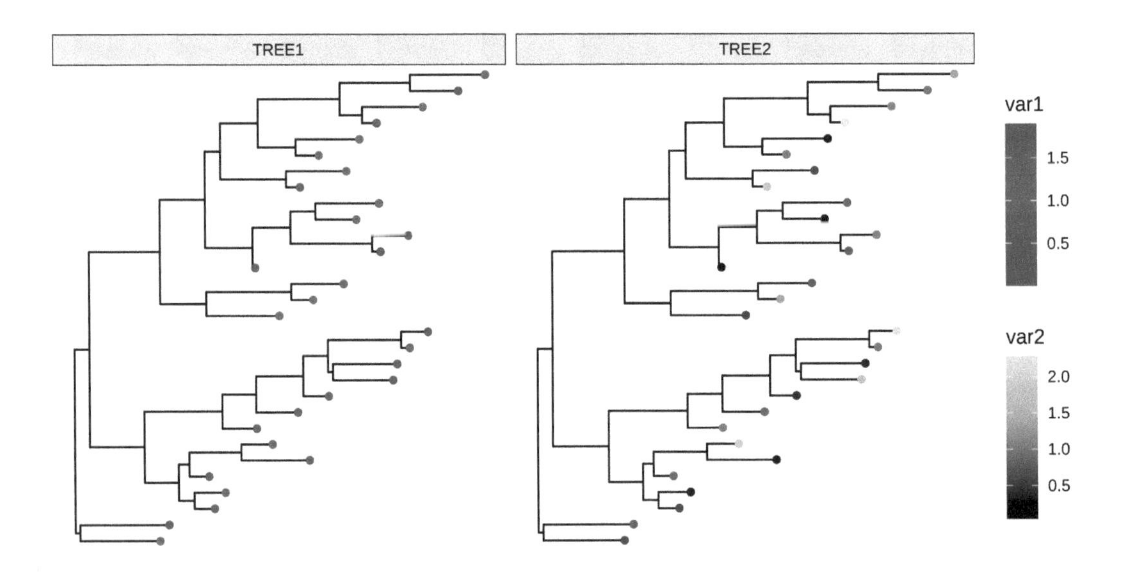

Figure 4.17: **Annotate one tree with values from different variables.** Using `subset` aesthetic mapping (as in TREE1 panel) or `td_filter()` (as in TREE2 panel) to filter variables to be displayed on the specific panel.

4.4.2 DensiTree

Another way to view the bootstrap trees is to merge them to form a density tree using `ggdensitree()` function (Figure 4.18). This will help us identify consensus and differences among a large set of trees. The trees will be stacked on top of each other and the structures of the trees will be rotated to ensure the consistency of the tip order. The tip order is determined by the `tip.order` parameter and by default (`tip.order = 'mode'`) the tips are ordered by the most commonly seen topology. The user can pass in a character vector to specify the tip order, or pass in an integer, N, to order the tips by the order of the tips in the Nth tree. Passing `mds` to `tip.order` will order the tips based on MDS (Multidimensional Scaling) of the path length between the tips, or passing `mds_dist` will order the tips based on MDS of the distance between the tips.

```
ggdensitree(btrees, alpha=.3, colour='steelblue') +
    geom_tiplab(size=3) + hexpand(.35)
```

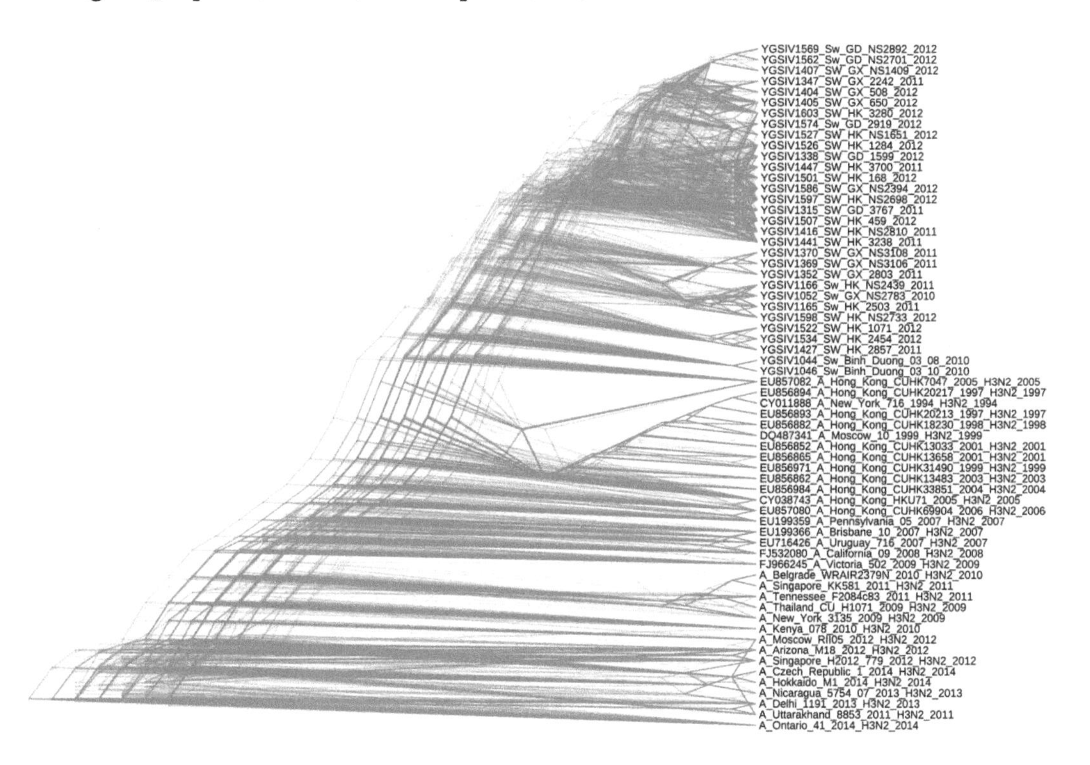

Figure 4.18: **DensiTree.** Trees are stacked on top of each other and the structures of the trees are rotated to ensure the consistency of the tip order.

4.5 Summary

Visualizing phylogenetic trees with **ggtree** is easy by using a single command `ggtree(tree)`. The **ggtree** package provides several geometric layers to display tree components such as tip labels, symbolic points for both external and internal nodes, root-edge, *etc.* Associated data can be used to rescale branch lengths, color the tree, and be displayed on the tree. All these can be done by the **ggplot2** grammar of graphic syntax, which makes it very easy to overlay layers and customize the tree graph (via **ggplot2** themes and scales). The **ggtree** package also provides several layers that are specifically designed for tree annotation which will be introduced in Chapter 5. The **ggtree** package makes the presentation of trees and associated data extremely easy. Simple graphs are easy to generate, while complex graphs are simply superimposed layers and are also easy to generate.

Chapter 5

Phylogenetic Tree Annotation

5.1 Visualizing and Annotating Tree Using Grammar of Graphics

The **ggtree** (Yu et al., 2017) is designed for a more general-purpose or a specific type of tree visualization and annotation. It supports the grammar of graphics implemented in **ggplot2** and users can freely visualize/annotate a tree by combining several annotation layers.

```
library(ggtree)
treetext = "(((ADH2:0.1[&&NHX:S=human], ADH1:0.11[&&NHX:S=human]):
0.05 [&&NHX:S=primates:D=Y:B=100],ADHY:
0.1[&&NHX:S=nematode],ADHX:0.12 [&&NHX:S=insect]):
0.1[&&NHX:S=metazoa:D=N],(ADH4:0.09[&&NHX:S=yeast],
ADH3:0.13[&&NHX:S=yeast], ADH2:0.12[&&NHX:S=yeast],
ADH1:0.11[&&NHX:S=yeast]):0.1[&&NHX:S=Fungi])[&&NHX:D=N];"
tree <- read.nhx(textConnection(treetext))
ggtree(tree) + geom_tiplab() +
  geom_label(aes(x=branch, label=S), fill='lightgreen') +
  geom_label(aes(label=D), fill='steelblue') +
  geom_text(aes(label=B), hjust=-.5)
```

Here, as an example, we visualized the tree with several layers to display annotation stored in NHX tags, including a layer of `geom_tiplab()` to display tip labels (gene name in this case), a layer using `geom_label()` to show species information (`S` tag) colored by light green, a layer of duplication event information (`D` tag) colored by steelblue and another layer using `geom_text()` to show bootstrap value (`B` tag).

Layers defined in **ggplot2** can be applied to **ggtree** directly as demonstrated in Figure 5.1 of using `geom_label()` and `geom_text()`. But **ggplot2** does not provide graphic layers that are specifically designed for phylogenetic tree annotation. For instance, layers for tip labels, tree branch scale legend, highlight, or labeling clade

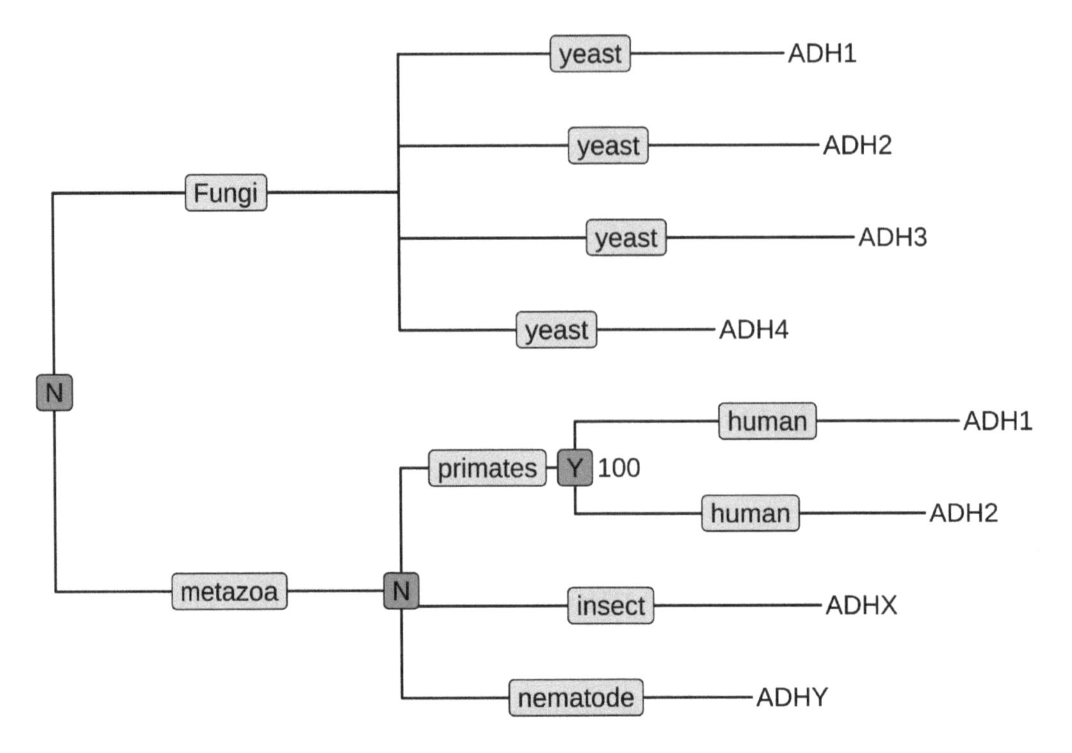

Figure 5.1: **Annotating tree using the grammar of graphics.** The NHX tree was annotated using the grammar of graphic syntax by combining different layers using the + operator. Species information was labeled in the middle of the branches. Duplication events were shown on the most recent common ancestor and clade bootstrap values were displayed near to it.

are all unavailable. To make tree annotation more flexible, several layers have been implemented in **ggtree** (Table 5.1), enabling different ways of annotation on various parts/components of a phylogenetic tree.

5.2 Layers for Tree Annotation

5.2.1 Colored strips

The **ggtree** (Yu et al., 2017) implements `geom_cladelab()` layer to annotate a selected clade with a bar indicating the clade with a corresponding label.

The `geom_cladelab()` layer accepts a selected internal node number and labels the corresponding clade automatically (Figure 5.2A). To get the internal node number, please refer to Chapter 2.

```
set.seed(2015-12-21)
tree <- rtree(30)
p <- ggtree(tree) + xlim(NA, 8)
```

Table 5.1: Geom layers defined in ggtree.

Layer	Description
geom_balance	Highlights the two direct descendant clades of an internal node
geom_cladelab	Annotates a clade with bar and text label (or image)
geom_facet	Plots associated data in a specific panel (facet) and aligns the plot with the tree
geom_hilight	Highlights selected clade with rectangular or round shape
geom_inset	Adds insets (subplots) to tree nodes
geom_label2	The modified version of geom_label, with subset aesthetic supported
geom_nodepoint	Annotates internal nodes with symbolic points
geom_point2	The modified version of geom_point, with subset aesthetic supported
geom_range	Bar layer to present uncertainty of evolutionary inference
geom_rootpoint	Annotates root node with symbolic point
geom_rootedge	Adds root-edge to a tree
geom_segment2	The modified version of geom_segment, with subset aesthetic supported
geom_strip	Annotates associated taxa with bar and (optional) text label
geom_taxalink	Links related taxa
geom_text2	The modified version of geom_text, with subset aesthetic supported
geom_tiplab	The layer of tip labels
geom_tippoint	Annotates external nodes with symbolic points
geom_tree	Tree structure layer, with multiple layouts supported
geom_treescale	Tree branch scale legend

```
p + geom_cladelab(node=45, label="test label") +
    geom_cladelab(node=34, label="another clade")
```

Users can set the parameter, `align = TRUE`, to align the clade label, `offset`, to adjust the position and color to set the color of the bar and label text, *etc.* (Figure 5.2B).

```
p + geom_cladelab(node=45, label="test label", align=TRUE,
                  offset = .2, textcolor='red', barcolor='red') +
    geom_cladelab(node=34, label="another clade", align=TRUE,
                  offset = .2, textcolor='blue', barcolor='blue')
```

Users can change the `angle` of the clade label text and relative position from text to bar via the parameter `offset.text`. The size of the bar and text can be changed via the parameters `barsize` and `fontsize`, respectively (Figure 5.2C).

```
p + geom_cladelab(node=45, label="test label", align=TRUE, angle=270,
            hjust='center', offset.text=.5, barsize=1.5, fontsize=8) +
    geom_cladelab(node=34, label="another clade", align=TRUE, angle=45)
```

Users can also use `geom_label()` to label the text and can set the background color by `fill` parameter (Figure 5.2D).

```
p + geom_cladelab(node=34, label="another clade", align=TRUE,
                  geom='label', fill='lightblue')
```

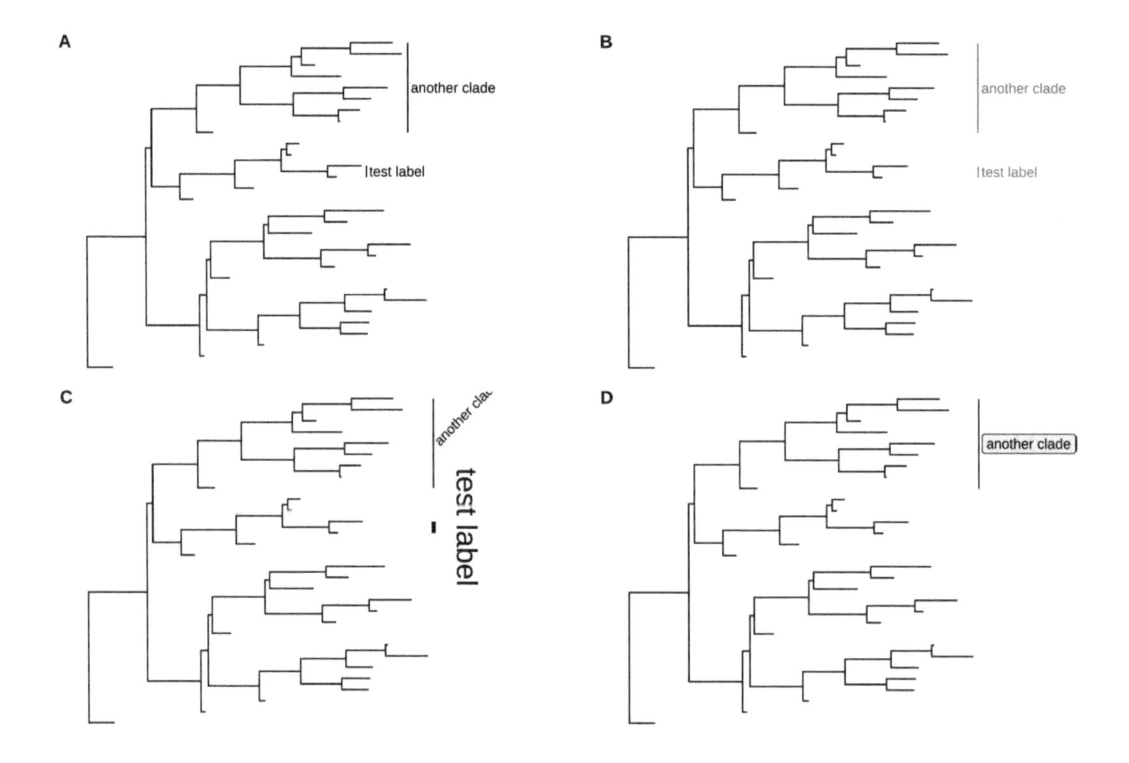

Figure 5.2: **Labeling clades.** Default (A); aligning and coloring clade bar and text (B); changing size and angle (C) and using `geom_label()` with background color in the text (D).

In addition, `geom_cladelab()` allows users to use the image or phylopic to annotate the clades, and supports using aesthetic mapping to automatically annotate the

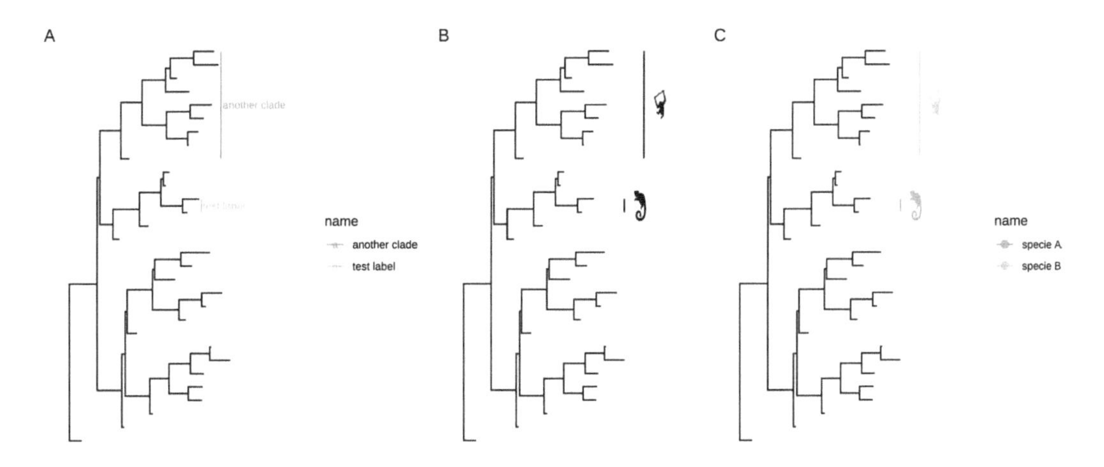

Figure 5.3: **Labeling clades using aesthetic mapping.** The geom_cladelab() layer allows users to use aesthetic mapping to annotate the clades (A); it supports using images or phylopic to annotate clades (B); mapping variable to change color or size of the text or image is also supported (C).

clade with bar and text label or image (*e.g.*, mapping variable to color the clade labels) (Figure 5.3).

```
dat <- data.frame(node = c(45, 34),
            name = c("test label", "another clade"))
# The node and label is required when geom="text"
## or geom="label" or geom="shadowtext".
p1 <- p + geom_cladelab(data = dat,
        mapping = aes(node = node, label = name, color = name),
        fontsize = 3)

dt <- data.frame(node = c(45, 34),
                image = c("7fb9bea8-e758-4986-afb2-95a2c3bf983d",
                          "0174801d-15a6-4668-bfe0-4c421fbe51e8"),
                name = c("specie A", "specie B"))

# when geom="phylopic" or geom="image", the image of aes is required.
p2 <- p + geom_cladelab(data = dt,
              mapping = aes(node = node, label = name, image = image),
              geom = "phylopic", imagecolor = "black",
              offset=1, offset.text=0.5)

# The color or size of image also can be mapped.
p3 <- p + geom_cladelab(data = dt,
              mapping = aes(node = node, label = name,
                          image = image, color = name),
```

```
          geom = "phylopic", offset = 1, offset.text=0.5)
```

The geom_cladelab() layer also supports unrooted tree layouts (Figure 5.4A).

```
ggtree(tree, layout="daylight") +
  geom_cladelab(node=35, label="test label", angle=0,
                fontsize=8, offset=.5, vjust=.5)  +
  geom_cladelab(node=55, label='another clade',
                angle=-95, hjust=.5, fontsize=8)
```

The geom_cladelab() is designed for labeling Monophyletic (Clade) while there are related taxa that do not form a clade. In **ggtree**, we provide another layer, geom_strip(), to add a strip/bar to indicate the association with an optional label for Polyphyletic or Paraphyletic (Figure 5.4B).

```
p + geom_tiplab() +
  geom_strip('t10', 't30', barsize=2, color='red',
          label="associated taxa", offset.text=.1) +
  geom_strip('t1', 't18', barsize=2, color='blue',
          label = "another label", offset.text=.1)
```

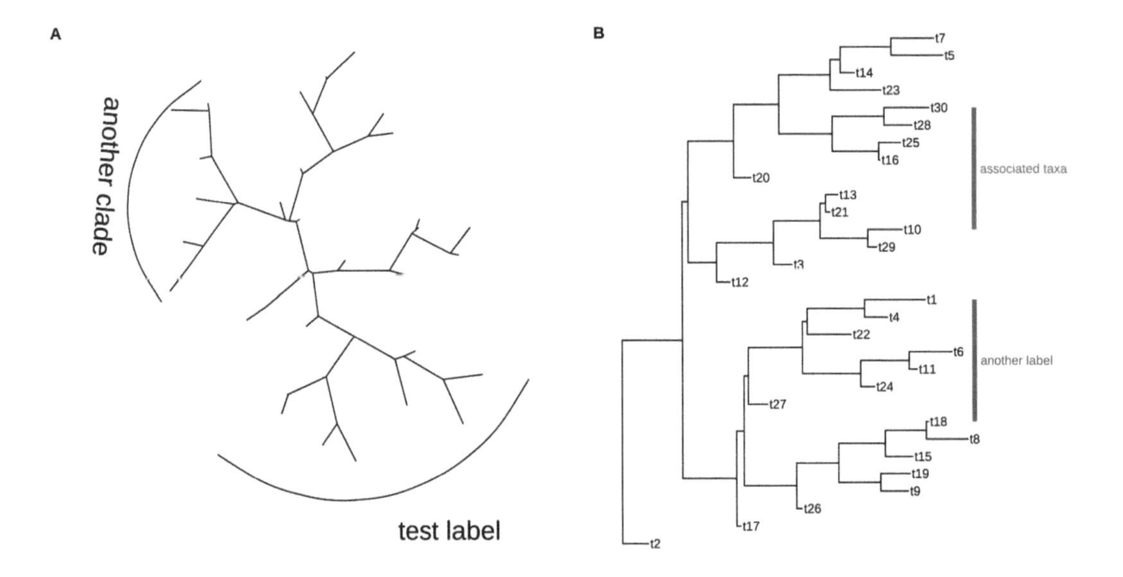

Figure 5.4: **Labeling associated taxa.** The geom_cladelab() is designed for labeling Monophyletic and supports unrooted layouts (A). The geom_strip() is designed for labeling all types of associated taxa, including Monophyletic, Polyphyletic, and Paraphyletic (B).

5.2.2 Highlight clades

The **ggtree** implements the `geom_hilight()` layer, which accepts an internal node number and adds a layer of a rectangle to highlight the selected clade (Figure 5.5)[1].

```
nwk <- system.file("extdata", "sample.nwk", package="treeio")
tree <- read.tree(nwk)
ggtree(tree) +
    geom_hilight(node=21, fill="steelblue", alpha=.6) +
    geom_hilight(node=17, fill="darkgreen", alpha=.6)

ggtree(tree, layout="circular") +
    geom_hilight(node=21, fill="steelblue", alpha=.6) +
    geom_hilight(node=23, fill="darkgreen", alpha=.6)
```

The `geom_hilight` layer also supports highlighting clades for unrooted layout trees with round ('encircle') or rectangular ('rect') shape (Figure 5.5C).

```
## type can be 'encircle' or 'rect'
pg + geom_hilight(node=55, linetype = 3) +
  geom_hilight(node=35, fill='darkgreen', type="rect")
```

Another way to highlight selected clades is by setting the clades with different colors and/or line types as demonstrated in Figure 6.2.

In addition to `geom_hilight()`, **ggtree** also implements `geom_balance()` which is designed to highlight neighboring subclades of a given internal node (Figure 5.5D).

```
ggtree(tree) +
  geom_balance(node=16, fill='steelblue', color='white', alpha=0.6,
    extend=1) +
  geom_balance(node=19, fill='darkgreen', color='white', alpha=0.6,
    extend=1)
```

The `geom_hilight()` layer supports using aesthetic mapping to automatically highlight clades as demonstrated in Figures 5.5E-F. For plot in Cartesian coordinates (*e.g.*, rectangular layout), the rectangle can be rounded (Figure 5.5E) or filled with gradient colors (Figure 5.5F).

```
## using external data
d <- data.frame(node=c(17, 21), type=c("A", "B"))
ggtree(tree) + geom_hilight(data=d, aes(node=node, fill=type),
                            type = "roundrect")

## using data stored in the tree object
x <- read.nhx(system.file("extdata/NHX/ADH.nhx", package="treeio"))
ggtree(x) + geom_hilight(mapping=aes(subset = node %in% c(10, 12),
```

[1] If you want to plot the tree above the highlighting area, visit FAQ for details.

```
                                            fill = S),
                                  type = "gradient", gradient.direction = 'rt',
                                  alpha = .8) +
  scale_fill_manual(values=c("steelblue", "darkgreen"))
```

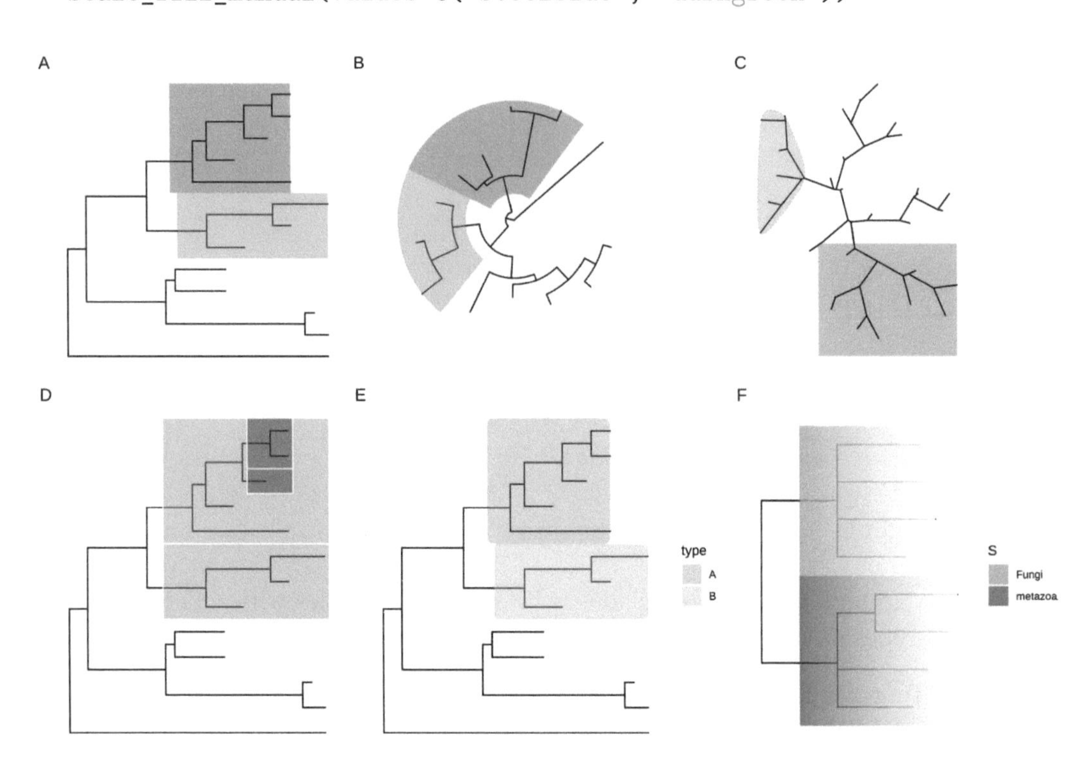

Figure 5.5: **Highlight selected clades.** Rectangular layout (A), circular/fan (B), and unrooted layouts. Highlight neighboring subclades simultaneously (D). Highlight selected clades using associated data (E and F).

5.2.3 Taxa connection

Some evolutionary events (*e.g.*, reassortment, horizontal gene transfer) cannot be modeled by a simple tree. The **ggtree** provides the `geom_taxalink()` layer that allows drawing straight or curved lines between any of two nodes in the tree, allowing it to represent evolutionary events by connecting taxa. It works with rectangular (Figure 5.6A), circular (Figure 5.6B), and inward circular (Figure 5.6C) layouts. The `geom_taxalink()` is not only useful for presenting evolutionary events, but it can also be used to combine evolutionary trees to present relationships or interactions between species (Xu, Dai, et al., 2021).

The `geom_taxalink()` layout supports aesthetic mapping, which requires a `data.frame` that stores association information with/without meta-data as input (Figure 5.6D).

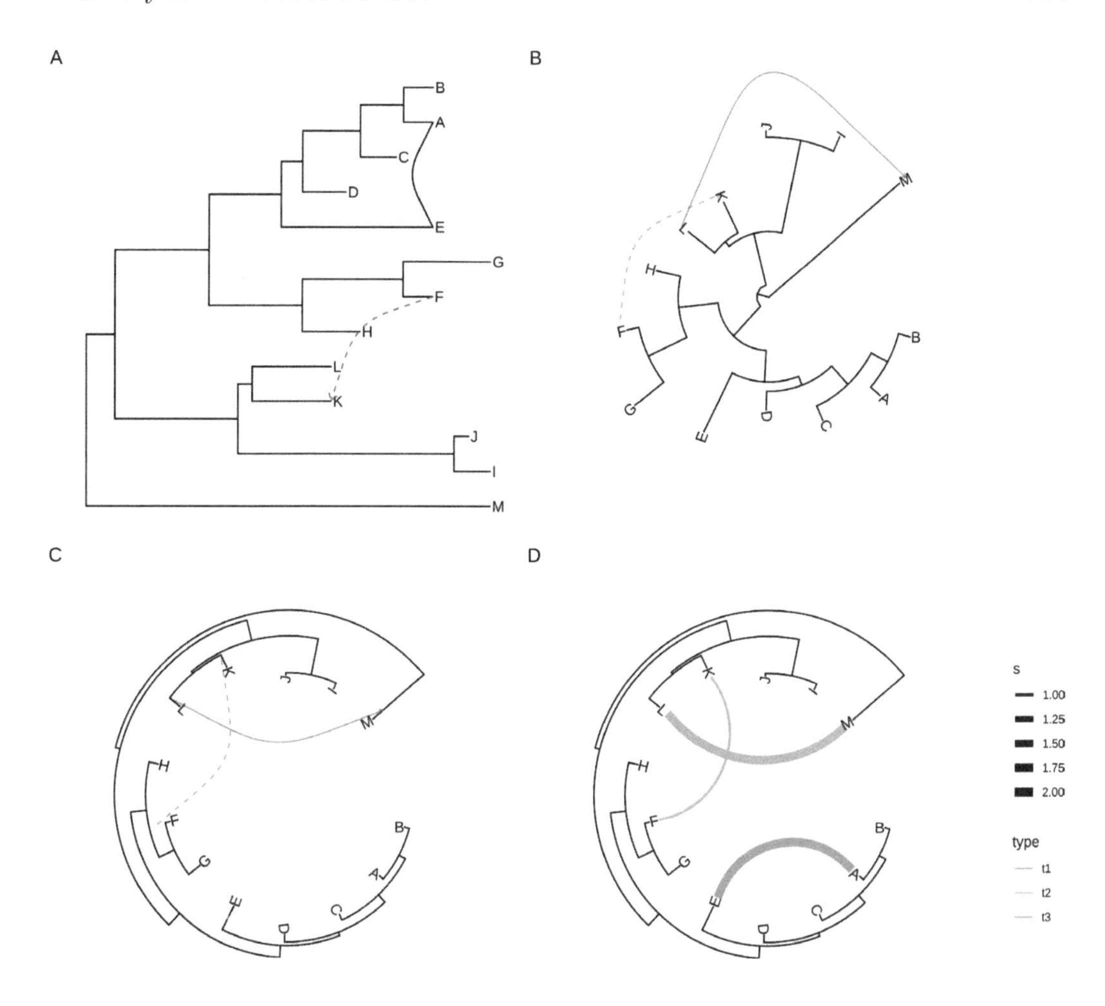

Figure 5.6: **Linking related taxa.** This can be used to indicate evolutionary events or relationships between species. Rectangular layout (A), circular layout (B), and inward circular layout (C and D). It supports aesthetic mapping to map variables to set line sizes and colors (D).

```
p1 <- ggtree(tree) + geom_tiplab() + geom_taxalink(taxa1='A', taxa2='E') +
  geom_taxalink(taxa1='F', taxa2='K', color='red', linetype = 'dashed',
    arrow=arrow(length=unit(0.02, "npc")))

p2 <- ggtree(tree, layout="circular") +
      geom_taxalink(taxa1='A', taxa2='E', color="grey", alpha=0.5,
                offset=0.05, arrow=arrow(length=unit(0.01, "npc"))) +
      geom_taxalink(taxa1='F', taxa2='K', color='red',
                linetype = 'dashed', alpha=0.5, offset=0.05,
                arrow=arrow(length=unit(0.01, "npc"))) +
      geom_taxalink(taxa1="L", taxa2="M", color="blue", alpha=0.5,
                offset=0.05, hratio=0.8,
                arrow=arrow(length=unit(0.01, "npc"))) +
```

```
    geom_tiplab()

# when the tree was created using reverse x,
# we can set outward to FALSE, which will generate the inward curve lines.
p3 <- ggtree(tree, layout="inward_circular", xlim=150) +
    geom_taxalink(taxa1='A', taxa2='E', color="grey", alpha=0.5,
                  offset=-0.2, outward=FALSE,
                  arrow=arrow(length=unit(0.01, "npc"))) +
    geom_taxalink(taxa1='F', taxa2='K', color='red', linetype = 'dashed',
                  alpha=0.5, offset=-0.2, outward=FALSE,
                  arrow=arrow(length=unit(0.01, "npc"))) +
    geom_taxalink(taxa1="L", taxa2="M", color="blue", alpha=0.5,
                  offset=-0.2, outward=FALSE,
                  arrow=arrow(length=unit(0.01, "npc"))) +
    geom_tiplab(hjust=1)

dat <- data.frame(from=c("A", "F", "L"),
                  to=c("E", "K", "M"),
                  h=c(1, 1, 0.1),
                  type=c("t1", "t2", "t3"),
                  s=c(2, 1, 2))
p4 <- ggtree(tree, layout="inward_circular", xlim=c(150, 0)) +
        geom_taxalink(data=dat,
                      mapping=aes(taxa1=from,
                                  taxa2=to,
                                  color=type,
                                  size=s),
                      ncp=10,
                      offset=0.15) +
        geom_tiplab(hjust=1) +
        scale_size_continuous(range=c(1,3))
plot_list(p1, p2, p3, p4, ncol=2, tag_levels='A')
```

5.2.4 Uncertainty of evolutionary inference

The `geom_range()` layer supports displaying interval (highest posterior density, confidence interval, range) as horizontal bars on tree nodes. The center of the interval will anchor to the corresponding node. The center by default is the mean value of the interval (Figure 5.7A). We can set the `center` to the estimated mean or median value (Figure 5.7B), or the observed value. As the tree branch and the interval may not be on the same scale, **ggtree** provides `scale_x_range` to add a second x-axis for the range (Figure 5.7C). Note that x-axis is disabled by the default theme, and we need to enable it if we want to display it (*e.g.*, using `theme_tree2()`).

```
file <- system.file("extdata/MEGA7", "mtCDNA_timetree.nex",
                    package = "treeio")
x <- read.mega(file)
p1 <- ggtree(x) + geom_range('reltime_0.95_CI', color='red', size=3,
```

```
                                 alpha=.3)
p2 <- ggtree(x) + geom_range('reltime_0.95_CI', color='red', size=3,
                                 alpha=.3, center='reltime')
p3 <- p2 + scale_x_range() + theme_tree2()
```

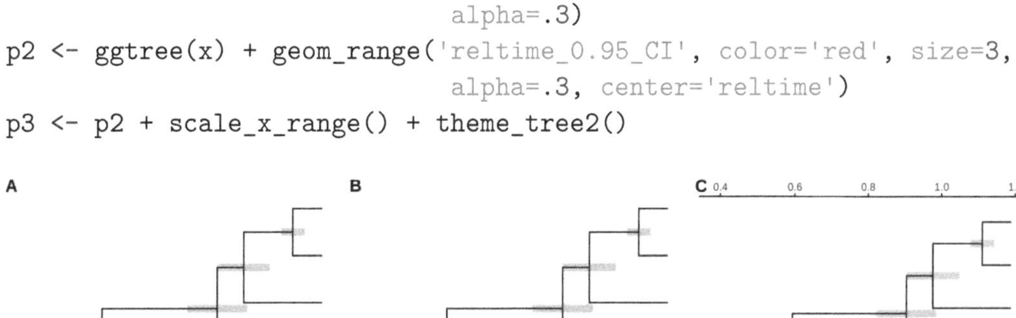
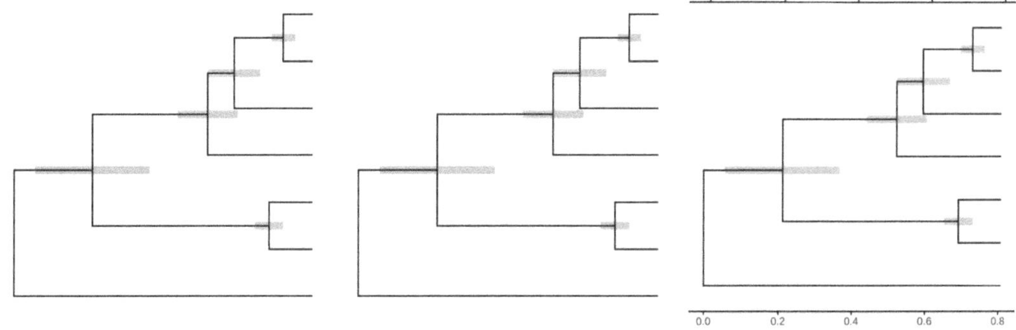

Figure 5.7: **Displaying uncertainty of evolutionary inference.** The center (mean value of the range (A) or estimated value (B)) is anchored to the tree nodes. A second x-axis was used for range scaling (C).

5.3 Tree Annotation with Output from Evolution Software

5.3.1 Tree annotation using data from evolutionary analysis software

Chapter 1 introduced using **treeio** package (Wang et al., 2020) to parse different tree formats and commonly used software outputs to obtain phylogeny-associated data. These imported data, as `S4` objects, can be visualized directly using **ggtree**. Figure 5.1 demonstrates a tree annotated using the information (species classification, duplication event, and bootstrap value) stored in the NHX file. **PHYLDOG** and `pkg_revbayes` output NHX files that can be parsed by **treeio** and visualized by **ggtree** with annotation using their inference data.

Furthermore, the evolutionary data from the inference of **BEAST**, **MrBayes**, and **RevBayes**, d_N/d_S values inferred by **CODEML**, ancestral sequences inferred by **HyPhy**, **CODEML**, or **BASEML** and short read placement by **EPA** and **PPLACER** can be used to annotate the tree directly.

```
file <- system.file("extdata/BEAST", "beast_mcc.tree", package="treeio")
beast <- read.beast(file)
ggtree(beast, aes(color=rate))  +
    geom_range(range='length_0.95_HPD', color='red', alpha=.6, size=2) +
    geom_nodelab(aes(x=branch, label=round(posterior, 2)), vjust=-.5,
                 size=3) +
```

```
    scale_color_continuous(low="darkgreen", high="red") +
    theme(legend.position=c(.1, .8))
```

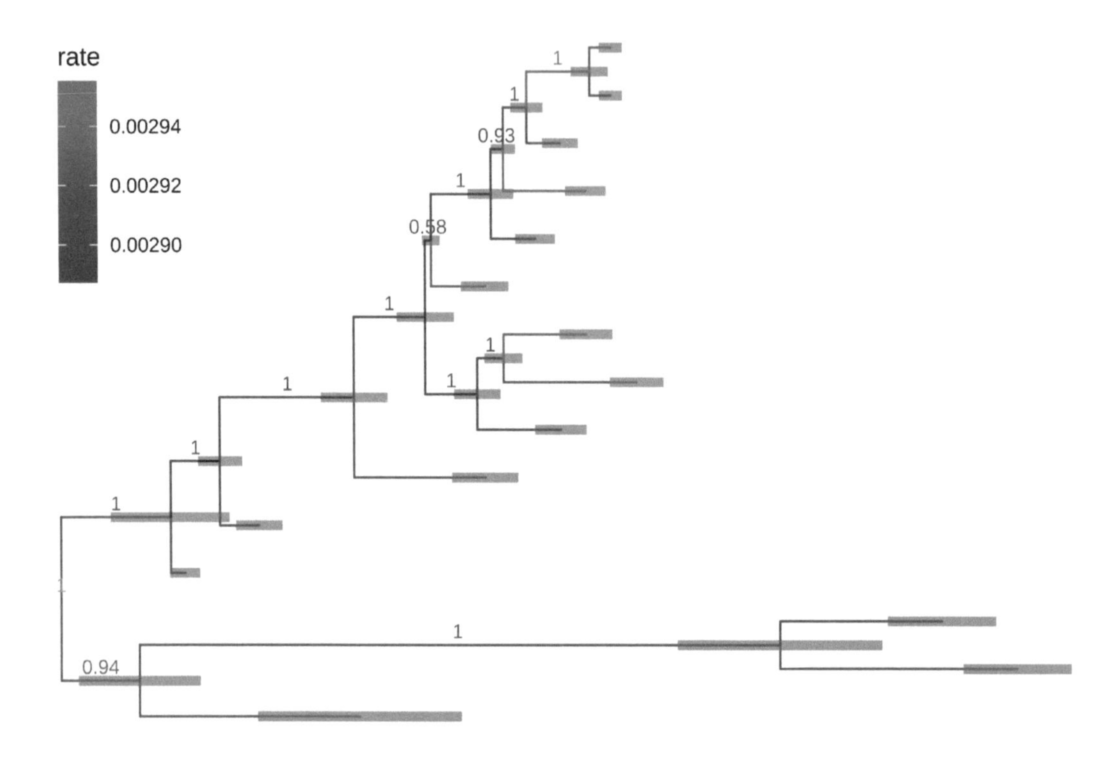

Figure 5.8: **Annotating BEAST tree with *length_95%_HPD* and posterior.** Branch length credible intervals (95% HPD) were displayed as red horizontal bars and clade posterior values were shown on the middle of branches.

In Figure 5.8, the tree was visualized and annotated with posterior >0.9 and demonstrated length uncertainty (95% Highest Posterior Density (HPD) interval).

Ancestral sequences inferred by **HyPhy** can be parsed using **treeio**, whereas the substitutions along each tree branch were automatically computed and stored inside the phylogenetic tree object (*i.e.*, `S4` object). The **ggtree** package can utilize this information stored in the object to annotate the tree, as demonstrated in Figure 5.9.

```
nwk <- system.file("extdata/HYPHY", "labelledtree.tree",
                   package="treeio")
ancseq <- system.file("extdata/HYPHY", "ancseq.nex",
                      package="treeio")
tipfas <- system.file("extdata", "pa.fas", package="treeio")
hy <- read.hyphy(nwk, ancseq, tipfas)
ggtree(hy) +
  geom_text(aes(x=branch, label=AA_subs), size=2,
            vjust=-.3, color="firebrick")
```

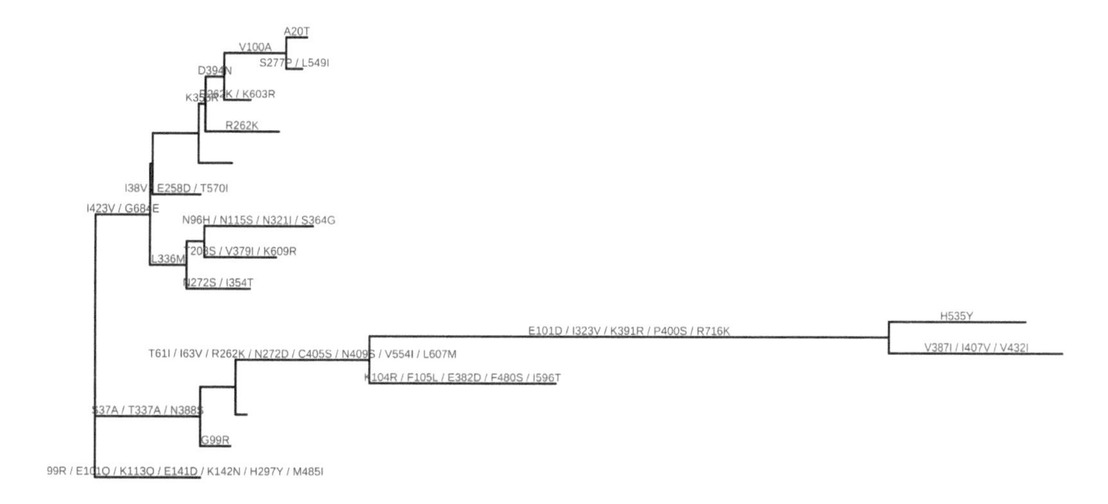

Figure 5.9: **Annotating tree with amino acid substitution determined by ancestral sequences inferred by HyPhy.** Amino acid substitutions were displayed in the middle of branches.

PAML's **BASEML** and **CODEML** can also be used to infer ancestral sequences, whereas **CODEML** can infer selection pressure. After parsing this information using **treeio**, **ggtree** can integrate this information into the same tree structure and be used for annotation as illustrated in Figure 5.10.

```
rstfile <- system.file("extdata/PAML_Codeml", "rst",
                       package="treeio")
mlcfile <- system.file("extdata/PAML_Codeml", "mlc",
                       package="treeio")
ml <- read.codeml(rstfile, mlcfile)
ggtree(ml, aes(color=dN_vs_dS), branch.length='dN_vs_dS') +
  scale_color_continuous(name='dN/dS', limits=c(0, 1.5),
                         oob=scales::squish,
                         low='darkgreen', high='red') +
  geom_text(aes(x=branch, label=AA_subs),
            vjust=-.5, color='steelblue', size=2) +
  theme_tree2(legend.position=c(.9, .3))
```

Not only all the tree data parsed by **treeio** can be used to visualize and annotate the phylogenetic tree using **ggtree**, but also other trees and tree-like objects defined in the R community are supported. The **ggtree** plays a unique role in the R ecosystem to facilitate phylogenetic analysis, and it can be easily integrated into other packages and pipelines. For more details of working with other tree-like structures, please refer to Chapter 9. In addition to direct support of tree objects, **ggtree** also allows users to plot a tree with different types of external data (see also Chapter 7 and (Yu et al., 2018)).

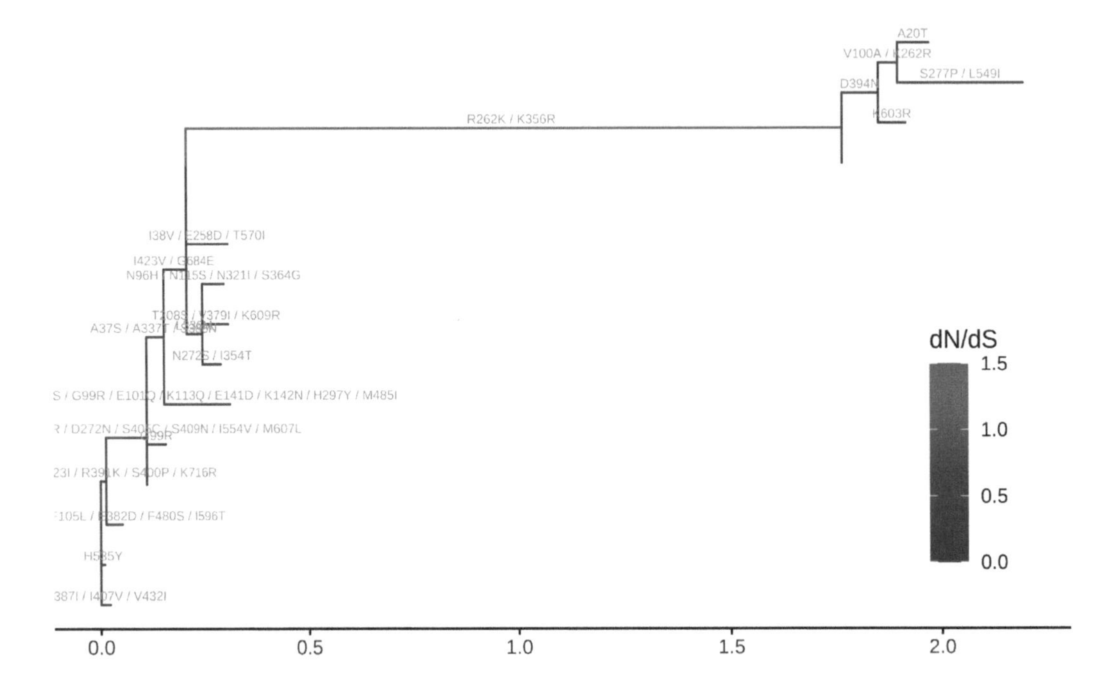

Figure 5.10: **Annotating tree with amino acid substitution and d_N/d_S inferred by CODEML.** Branches were rescaled and colored by d_N/d_S values, and amino acid substitutions were displayed on the middle of branches.

5.4 Summary

The **ggtree** package implements the grammar of graphics for annotating phylogenetic trees. Users can use the **ggplot2** syntax to combine different annotation layers to produce complex tree annotation. If you are familiar with **ggplot2**, tree annotation with a high level of customization can be intuitive and flexible using **ggtree**. The **ggtree** can collect information in the `treedata` object or link external data to the structure of the tree. This will enable us to use the phylogenetic tree for data integration analysis and comparative studies, and will greatly expand the application of the phylogenetic tree in different fields.

Chapter 6

Visual Exploration of Phylogenetic Trees

The **ggtree** (Yu et al., 2017) supports many ways of manipulating the tree visually, including viewing selected clade to explore large tree (Figure 6.1), taxa clustering (Figure 6.5), rotating clade or tree (Figure 6.6B and 6.8), zoom out or collapsing clades (Figure 6.3A and 6.2), *etc.*. Details of the tree manipulation functions are summarized in Table 6.1.

Table 6.1: Tree manipulation functions.

Function	Description
collapse	Collapse a selecting clade
expand	Expand collapsed clade
flip	Exchange position of 2 clades that share a parent node
groupClade	Grouping clades
groupOTU	Grouping OTUs by tracing back to the most recent common ancestor
identify	Interactive tree manipulation
rotate	Rotating a selected clade by 180 degrees
rotate_tree	Rotating circular layout tree by a specific angle
scaleClade	Zoom in or zoom out selecting clade
open_tree	Convert a tree to fan layout by specific open angle

6.1 Viewing Selected Clade

A clade is a monophyletic group that contains a single ancestor and all of its descendants. We can visualize a specifically selected clade via the `viewClade()` function as demonstrated in Figure 6.1B. Another solution is to extract the selected clade as a new tree object as described in session 2.5. These functions are developed to help users explore a large tree.

```
library(ggtree)
nwk <- system.file("extdata", "sample.nwk", package="treeio")
tree <- read.tree(nwk)
p <- ggtree(tree) + geom_tiplab()
viewClade(p, MRCA(p, "I", "L"))
```

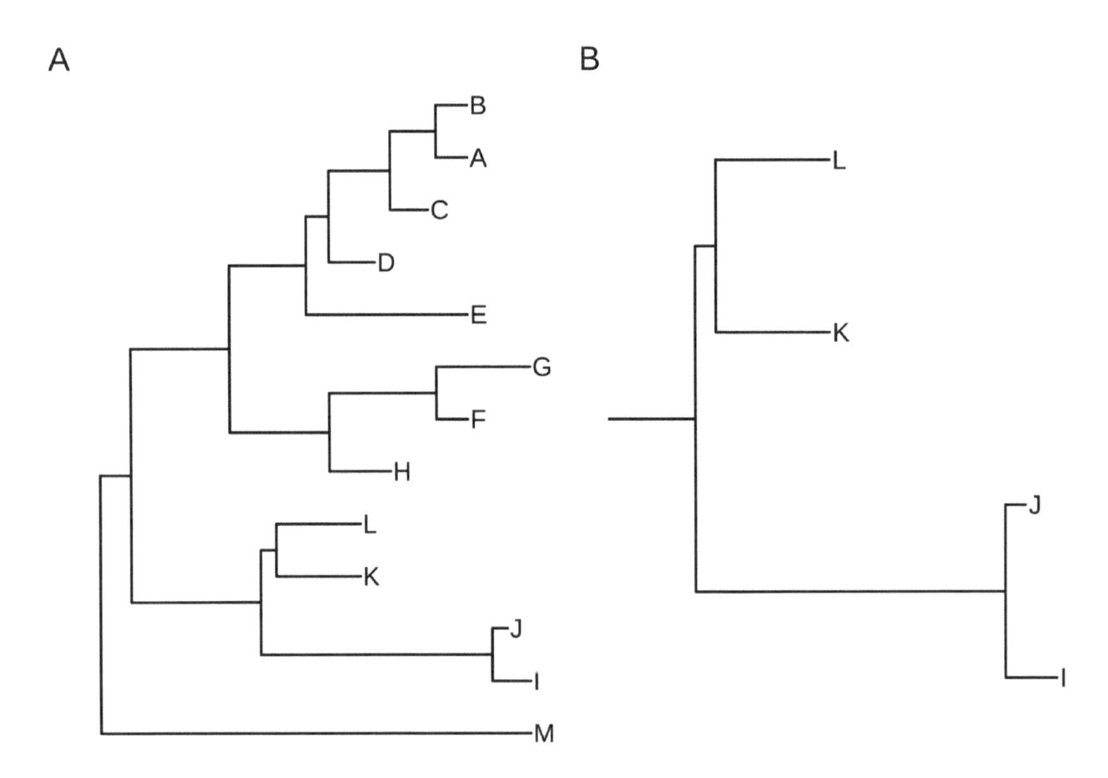

Figure 6.1: **Viewing a selected clade of a tree.** An example tree used to demonstrate how **ggtree** supports exploring or manipulating phylogenetic tree visually (A). The **ggtree** supports visualizing selected clade (B). A clade can be selected by specifying a node number or determined by the most recent common ancestor of selected tips.

Some of the functions, *e.g.*, `viewClade()`, work with clade and accept a parameter of an internal node number. To get the internal node number, users can use the `MRCA()` function (as in Figure 6.1) by providing two taxa names. The function will return the node number of input taxa's most recent common ancestor (MRCA). It works with a tree and graphic (*i.e.*, the `ggtree()` output) object. The **tidytree** package also provides an `MRCA()` method to extract information from the MRCA node (see details in session 2.1.3).

6.2 Scaling Selected Clade

The **ggtree** provides another option to zoom out (or compress) selected clades via the `scaleClade()` function. In this way, we retain the topology and branch lengths of compressed clades. This helps to save the space to highlight those clades of primary interest in the study.

```
tree2 <- groupClade(tree, c(17, 21))
p <- ggtree(tree2, aes(color=group)) + theme(legend.position='none') +
  scale_color_manual(values=c("black", "firebrick", "steelblue"))
scaleClade(p, node=17, scale=.1)
```

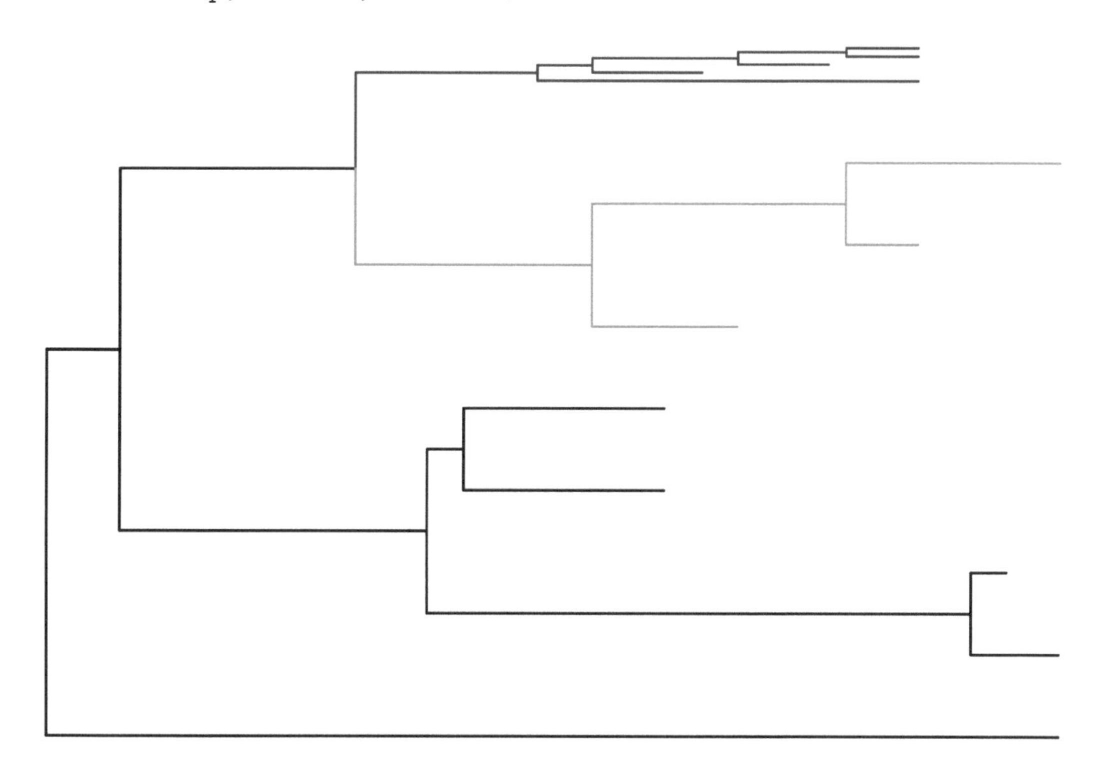

Figure 6.2: **Scaling selected clade.** Clades can be zoomed in (if `scale > 1`) to highlight or zoomed out to save space.

If users want to emphasize important clades, they can use the `scaleClade()` function by passing a numeric value larger than 1 to the `scale` parameter. Then the selected clade will be zoomed in. Users can also use the `groupClade()` function to assign selected clades with different clade IDs which can be used to color these clades with different colors as shown in Figure 6.2.

6.3 Collapsing and Expanding Clade

It is a common practice to prune or collapse clades so that certain aspects of a tree can be emphasized. The **ggtree** supports collapsing selected clades using the `collapse()` function as shown in Figure 6.3A.

```
p2 <- p %>% collapse(node=21) +
  geom_point2(aes(subset=(node==21)), shape=21, size=5, fill='green')
p2 <- collapse(p2, node=23) +
  geom_point2(aes(subset=(node==23)), shape=23, size=5, fill='red')
print(p2)
expand(p2, node=23) %>% expand(node=21)
```

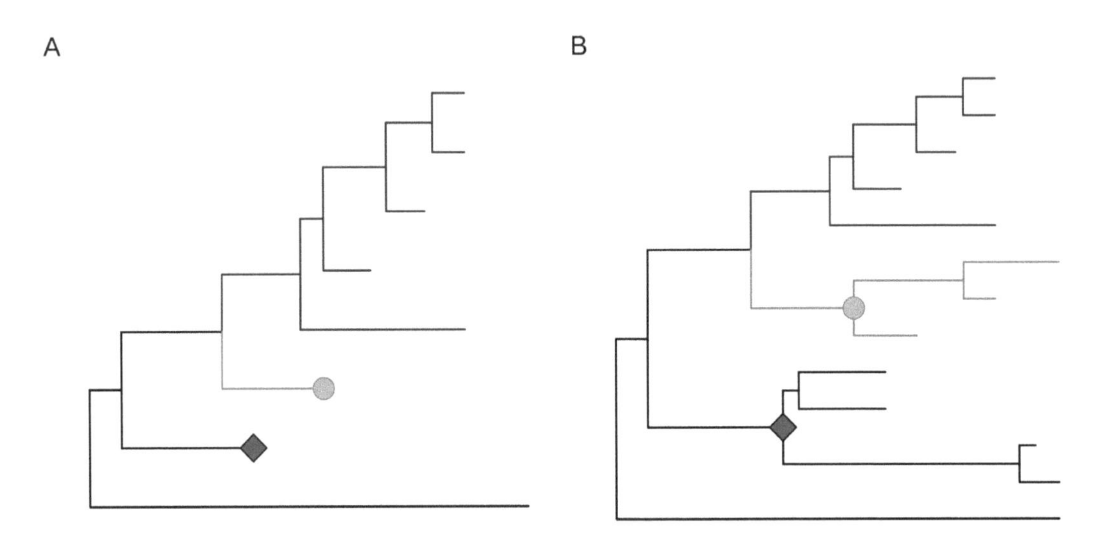

Figure 6.3: **Collapsing selected clades and expanding collapsed clades.** Clades can be selected to collapse (A) and the collapsed clades can be expanded back (B) if necessary as **ggtree** stored all information of species relationships. Green and red symbols were displayed on the tree to indicate the collapsed clades.

Here two clades were collapsed and labeled by the green circle and red square symbolic points. Collapsing is a common strategy to collapse clades that are too large for displaying in full or are not the primary interest of the study. In **ggtree**, we can expand (*i.e.*, uncollapse) the collapsed branches back with the `expand()` function to show details of species relationships as demonstrated in Figure 6.3B.

Triangles are often used to represent the collapsed clade and **ggtree** also supports it. The `collapse()` function provides a "mode" parameter, which by default is "none" and the selected clade was collapsed as a "tip". Users can specify the `mode` to "max" that uses the farthest tip (Figure 6.4A), "min" that uses the closest tip (Figure 6.4B), and "mixed" that uses both (Figure 6.4C).

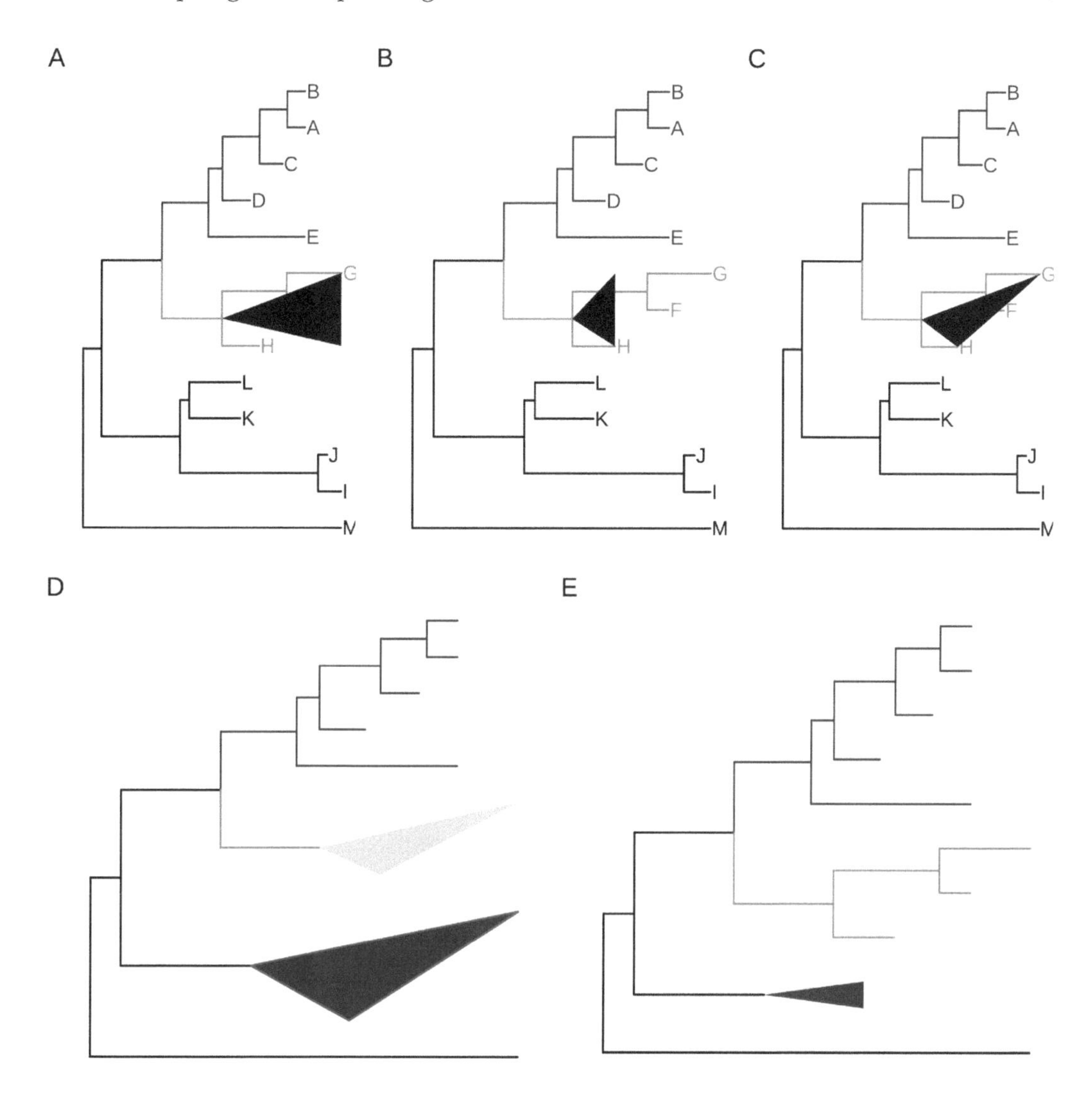

Figure 6.4: **Collapse clade as a triangle.** 'max' takes the position of most distant tip (A). 'min' takes the position of closest tip (B). 'mixed' takes the positions of both closest and distant tips (C), which looks more like the shape of the clade. Set color, fill, and alpha of the triangles (D). Combine with `scaleClade` to zoom out the triangle to save space (E).

```
p2 <- p + geom_tiplab()
node <- 21
collapse(p2, node, 'max') %>% expand(node)
collapse(p2, node, 'min') %>% expand(node)
collapse(p2, node, 'mixed') %>% expand(node)
```

We can pass additional parameters to set the color and transparency of the triangles (Figure 6.4D).

```
collapse(p, 21, 'mixed', fill='steelblue', alpha=.4) %>%
  collapse(23, 'mixed', fill='firebrick', color='blue')
```

We can combine scaleClade with `collapse` to zoom in/out of the triangles (Figure 6.4E).

```
scaleClade(p, 23, .2) %>% collapse(23, 'min', fill="darkgreen")
```

6.4 Grouping Taxa

The `groupClade()` function assigns the branches and nodes under different clades into different groups. It accepts an internal node or a vector of internal nodes to cluster clade/clades.

Similarly, the `groupOTU()` function assigns branches and nodes to different groups based on user-specified groups of operational taxonomic units (OTUs) that are not necessarily within a clade but can be monophyletic (clade), polyphyletic or paraphyletic. It accepts a vector of OTUs (taxa name) or a list of OTUs and will trace back from OTUs to their most recent common ancestor (MRCA) and cluster them together as demonstrated in Figure 6.5.

A phylogenetic tree can be annotated by mapping different line types, sizes, colors, or shapes of the branches or nodes that have been assigned to different groups.

```
data(iris)
rn <- paste0(iris[,5], "_", 1:150)
rownames(iris) <- rn
d_iris <- dist(iris[,-5], method="man")

tree_iris <- ape::bionj(d_iris)
grp <- list(setosa     = rn[1:50],
            versicolor = rn[51:100],
            virginica  = rn[101:150])

p_iris <- ggtree(tree_iris, layout = 'circular', branch.length='none')
groupOTU(p_iris, grp, 'Species') + aes(color=Species) +
  theme(legend.position="right")
```

We can group taxa at the tree level. The following code will produce an identical figure of Figure 6.5 (see more details described in session 2.2.3).

```
tree_iris <- groupOTU(tree_iris, grp, "Species")
ggtree(tree_iris, aes(color=Species), layout = 'circular',
        branch.length = 'none') +
  theme(legend.position="right")
```

Figure 6.5: **Grouping OTUs.** OTU clustering based on their relationships. Selected OTUs and their ancestors up to the MRCA will be clustered together.

6.5 Exploring Tree Structure

To facilitate exploring the tree structure, **ggtree** supports rotating selected clade by 180 degrees using the `rotate()` function (Figure 6.6B). Position of immediate descendant clades of the internal node can be exchanged via `flip()` function (Figure 6.6C).

```
p1 <- p + geom_point2(aes(subset=node==16), color='darkgreen', size=5)
p2 <- rotate(p1, 16)
flip(p2, 17, 21)
```

Most of the tree manipulation functions are working on clades, while **ggtree** also provides functions to manipulate a tree, including `open_tree()` to transform a tree in either rectangular or circular layout to the fan layout, and `rotate_tree()` function to rotate a tree for specific angle in both circular or fan layouts, as demonstrated in Figures 6.7 and 6.8.

```
p3 <- open_tree(p, 180) + geom_tiplab()
print(p3)

rotate_tree(p3, 180)
```

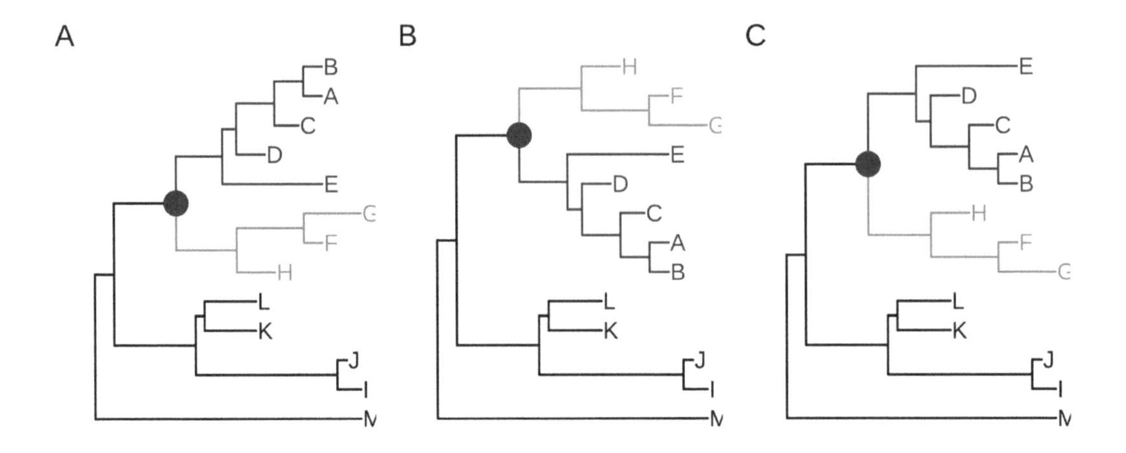

Figure 6.6: **Exploring tree structure.** A clade (indicated by a dark green circle) in a tree (A) can be rotated by 180° (B) and the positions of its immediate descendant clades (colored by blue and red) can be exchanged (C).

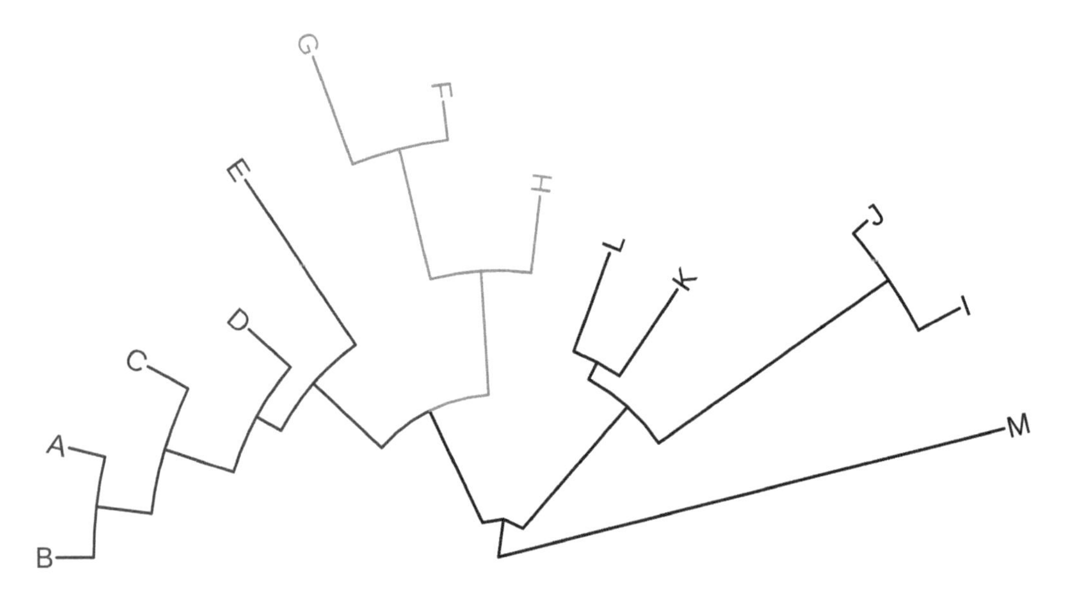

Figure 6.7: **Transforming a tree to fan layout.** A tree can be transformed to a fan layout by `open_tree` with a specific `angle`.

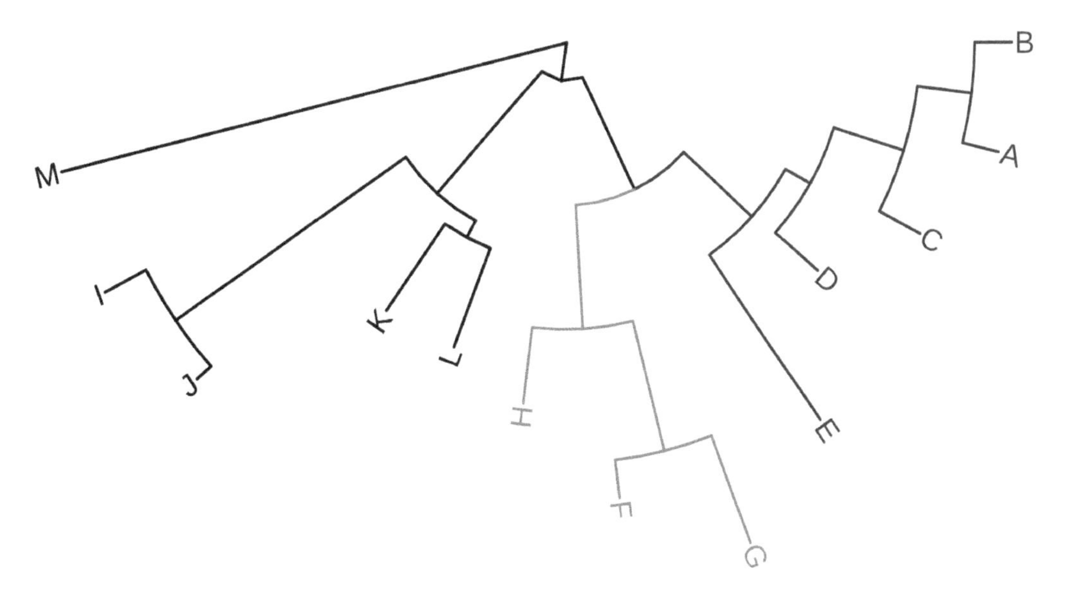

Figure 6.8: **Rotating tree.** A circular/fan layout tree can be rotated by any specific `angle`.

The following example rotates four selected clades (Figure 6.9). It is easy to traverse all the internal nodes and rotate them one-by-one.

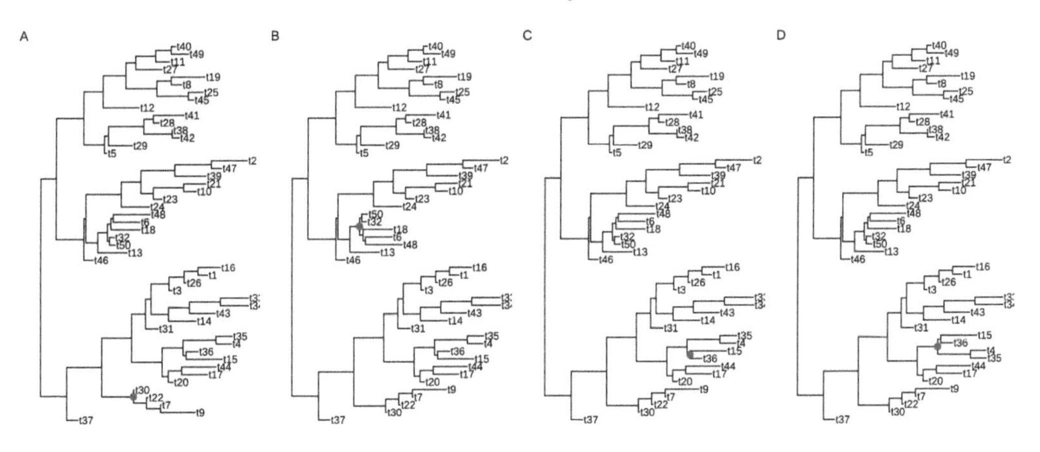

Figure 6.9: **Rotate selected clades.** Four clades were randomly selected to rotate (indicated by the red symbol).

```
set.seed(2016-05-29)
x <- rtree(50)
p <- ggtree(x) + geom_tiplab()

## nn <- unique(reorder(x, 'postorder')$edge[,1])
## to traverse all the internal nodes

nn <- sample(unique(reorder(x, 'postorder')$edge[,1]), 4)
```

```
pp <- lapply(nn, function(n) {
    p <- rotate(p, n)
    p + geom_point2(aes(subset=(node == n)), color='red', size=3)
})

plot_list(gglist=pp, tag_levels='A', nrow=1)
```

Figure 6.10 demonstrates the usage of `open_tree()` with different open angles.

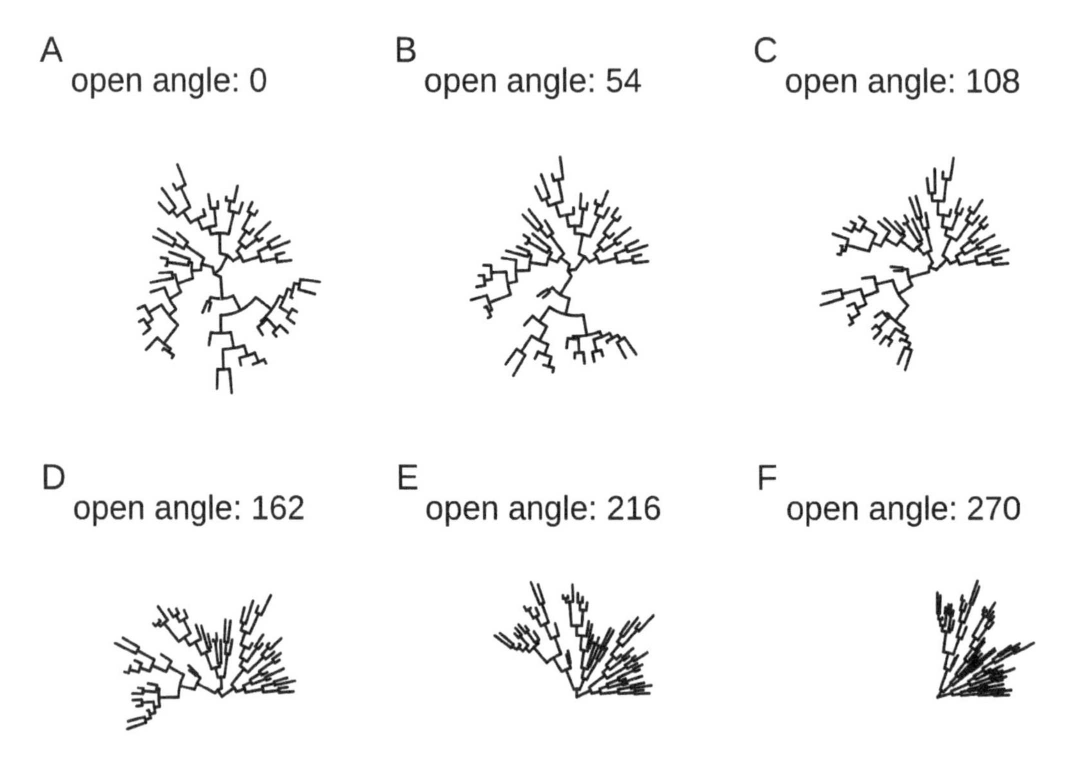

Figure 6.10: **Open tree with different angles.**

```
set.seed(123)
tr <- rtree(50)
p <- ggtree(tr, layout='circular')
angles <- seq(0, 270, length.out=6)

pp <- lapply(angles, function(angle) {
  open_tree(p, angle=angle) + ggtitle(paste("open angle:", angle))
})

plot_list(gglist=pp, tag_levels='A', nrow=2)
```

```
set.seed(123)
tr <- rtree(50)
p <- ggtree(tr, layout='circular')
angles <- seq(0, 270, length.out=6)

pp <- lapply(angles, function(angle) {
  open_tree(p, angle=angle) + ggtitle(paste("open angle:", angle))
})

g <- plot_list(gglist=pp, tag_levels='A', nrow=2)
ggplotify::as.ggplot(g, vjust=-.1,scale=1.1)
```

Figure 6.11 illustrates a rotating tree with different angles.

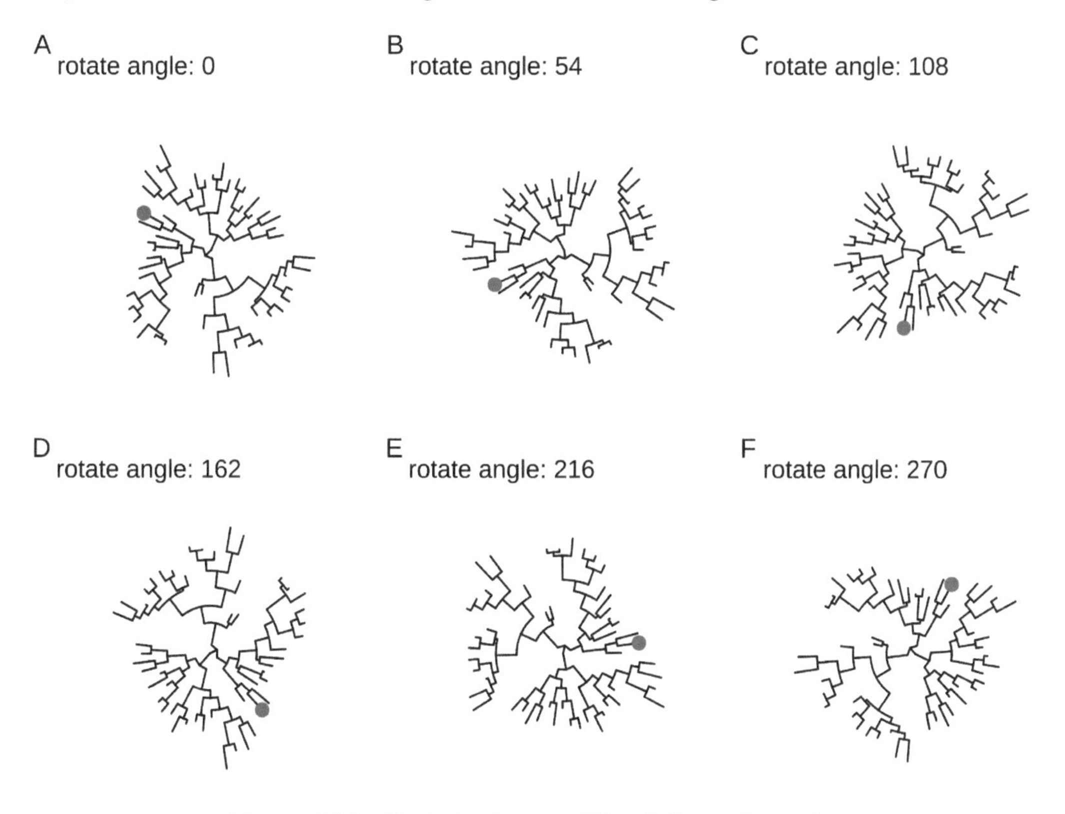

Figure 6.11: **Rotate tree with different angles.**

Interactive tree manipulation is also possible via the `identify()` method (see details described in Chapter 12).

6.6 Summary

A good visualization tool can not only help users to present the data, but it should also be able to help users to explore the data. Data exploration can allow users to better understand the data and also help discover hidden patterns. The **ggtree** provides a set of functions to allow visually manipulating phylogenetic trees and exploring tree structures with associated data. Exploring data in the evolutionary context may help discover new systematic patterns and generate new hypotheses.

Chapter 7

Plotting Tree with Data

Integrating user data to annotate a phylogenetic tree can be done at different levels. The **treeio** package (Wang et al., 2020) implements `full_join()` methods to combine tree data to phylogenetic tree object. The **tidytree** package supports linking tree data to phylogeny using tidyverse verbs (see also Chapter 2). The **ggtree** package (Yu et al., 2018) supports mapping external data to phylogeny for visualization and annotation on the fly. Although the feature of linking external data is overlapping among these packages, they have different application scopes. For example, in addition to the `treedata` object, **ggtree** also supports several other tree objects (see Chapter 9), including `phylo4d`, `phyloseq`, and `obkData` that were designed to contain domain-specific data. The design of these objects did not consider supporting linking external data to the object (it can not be done at the tree object level). We can visualize trees from these objects using **ggtree** and link external data at the visualization level (Yu et al., 2018).

The **ggtree** package provides two general methods for mapping and visualizing associated external data on phylogenies. Method 1 allows external data to be mapped on the tree structure and used as visual characteristics in the tree and data visualization. Method 2 plots the data with the tree side-by-side using different geometric functions after reordering the data based on the tree structure. These two methods integrate data with phylogeny for further exploration and comparison in the evolutionary biology context. The **ggtreeExtra** provides a better implementation of the Method 2 proposed in **ggtree** (see also Chapter 10) and works with both rectangular and circular layouts (Xu, Dai, et al., 2021).

7.1 Mapping Data to The tree Structure

In **ggtree**, we implemented an operator, `%<+%`, to attach annotation data to a `ggtree` graphic object. Any data that contains a column of "node" or the first column of taxa labels can be integrated using the `%<+%` operator. Multiple datasets can be attached progressively. When the data are attached, all the information stored in the data serves as numerical/categorical node attributes and can be directly used to visualize the tree by scaling the attributes as different colors or line sizes, labeling the tree using the original values of the attributes or parsing them as math expression, emoji or silhouette image. The following example uses the `%<+%` operator to integrate taxon (`df_tip_data`) and internal node (`df_inode_data`) information and map the data to different colors or shapes of symbolic points and labels (Figure 7.1). The tip data contains `imageURL` that links to online figures of the species, which can be parsed and used as tip labels in **ggtree** (see Chapter 8).

```
library(ggimage)
library(ggtree)
library(TDbook)

# load `tree_boots`, `df_tip_data`, and `df_inode_data` from 'TDbook'
p <- ggtree(tree_boots) %<+% df_tip_data + xlim(-.1, 4)
p2 <- p + geom_tiplab(offset = .6, hjust = .5) +
    geom_tippoint(aes(shape = trophic_habit, color = trophic_habit,
                size = mass_in_kg)) +
    theme(legend.position = "right") +
    scale_size_continuous(range = c(3, 10))

p2 %<+% df_inode_data +
 geom_label(aes(label = vernacularName.y, fill = posterior)) +
 scale_fill_gradientn(colors = RColorBrewer::brewer.pal(3, "YlGnBu"))
```

Although the data integrated by the `%<+%` operator in **ggtree** is for tree visualization, the data attached to the `ggtree` graphic object can be converted to `treedata` object that contains the tree and the attached data (see session 7.5).

7.2 Aligning Graph to the Tree Based on the Tree Structure

For associating phylogenetic tree with different types of plot produced by user's data, **ggtree** provides `geom_facet()` layer and `facet_plot()` function which accept an input `data.frame` and a `geom` layer to draw the input data. The data will be displayed in an additional panel of the plot. The `geom_facet()` (or `facet_plot`) is a general solution for linking the graphic layer to a tree. The function internally re-orders the input data based on the tree structure and visualizes the data at the specific panel by the geometric layer. Users are free to visualize several panels to plot different types of data as demonstrated in Figure 9.4 and to use different geometric layers to plot the same dataset (Figure 13.1) or different datasets on the same panel.

The `geom_facet()` is designed to work with most of the `geom` layers defined in **ggplot2** and other **ggplot2**-based packages. A list of the geometric layers that work seamlessly with `geom_facet()` and `facet_plot()` can be found in Table 1. As the

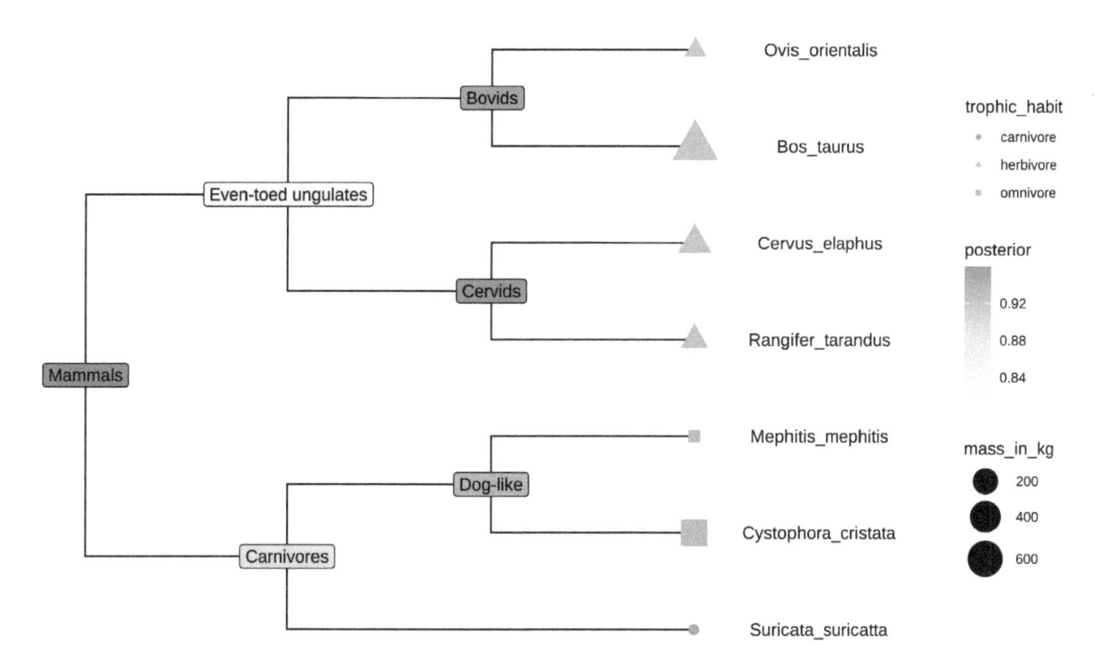

Figure 7.1: **Example of attaching multiple datasets**. External datasets including tip data (e.g., trophic habit and body weight) and node data (e.g., clade posterior and vernacular name) were attached to the `ggtree` graphic via the `%<+%` operator and the data was used to annotate the tree.

ggplot2 community keeps expanding and more `geom` layers will be implemented in either **ggplot2** or other extensions, `geom_facet()` and `facet_plot()` will gain more power to present data in the future. Note that different `geom` layers can be combined to present data on the same panel and the combinations of different `geom` layers create the possibility to present more complex data with phylogeny (see also Figures 13.1 and 13.4). Users can progressively add multiple panels to present and compare different datasets in the evolutionary context (Figure 7.2). Detailed descriptions can be found in the supplemental file of (Yu et al., 2018).

```
library(ggtree)
library(TDbook)

## load `tree_nwk`, `df_info`, `df_alleles`, and `df_bar_data`
## from 'TDbook'
tree <- tree_nwk
snps <- df_alleles
snps_strainCols <- snps[1,]
snps<-snps[-1,] # drop strain names
colnames(snps) <- snps_strainCols

gapChar <- "?"
```

```
snp <- t(snps)
lsnp <- apply(snp, 1, function(x) {
        x != snp[1,] & x != gapChar & snp[1,] != gapChar
    })
lsnp <- as.data.frame(lsnp)
lsnp$pos <- as.numeric(rownames(lsnp))
lsnp <- tidyr::gather(lsnp, name, value, -pos)
snp_data <- lsnp[lsnp$value, c("name", "pos")]

## visualize the tree
p <- ggtree(tree)

## attach the sampling information data set
## and add symbols colored by location
p <- p %<+% df_info + geom_tippoint(aes(color=location))

## visualize SNP and Trait data using dot and bar charts,
## and align them based on tree structure
p + geom_facet(panel = "SNP", data = snp_data, geom = geom_point,
               mapping=aes(x = pos, color = location), shape = '|') +
    geom_facet(panel = "Trait", data = df_bar_data, geom = geom_col,
                aes(x = dummy_bar_value, color = location,
                fill = location), orientation = 'y', width = .6) +
    theme_tree2(legend.position=c(.05, .85))
```

Companion functions to adjust panel widths and rename panel names are described in session 12.1. Removing the panel name is also possible and an example was presented in Figure 13.4. We can also use **aplot** or **patchwork** to create composite plots as described in session 7.5.

The `geom_facet()` (or `facet_plot()`) internally used `ggplot2::facet_grid()` and only works with Cartesian coordinate system. To align the graph to the tree for the polar system (e.g., for circular or fan layouts), we developed another Bioconductor package, **ggtreeExtra**. The **ggtreeExtra** package provides the `geom_fruit()` layer that works similar to `geom_facet()` (details described in Chapter 10). The `geom_fruit()` is a better implementation of the Method 2 proposed in (Yu et al., 2018).

7.3 Visualize a Tree with an Associated Matrix

The `gheatmap()` function is designed to visualize the phylogenetic tree with a heatmap of an associated matrix (either numerical or categorical). The `geom_facet()` layer is a general solution for plotting data with the tree, including heatmap. The `gheatmap()` function is specifically designed for plotting heatmap with a tree and provides a shortcut for handling column labels and color palettes.

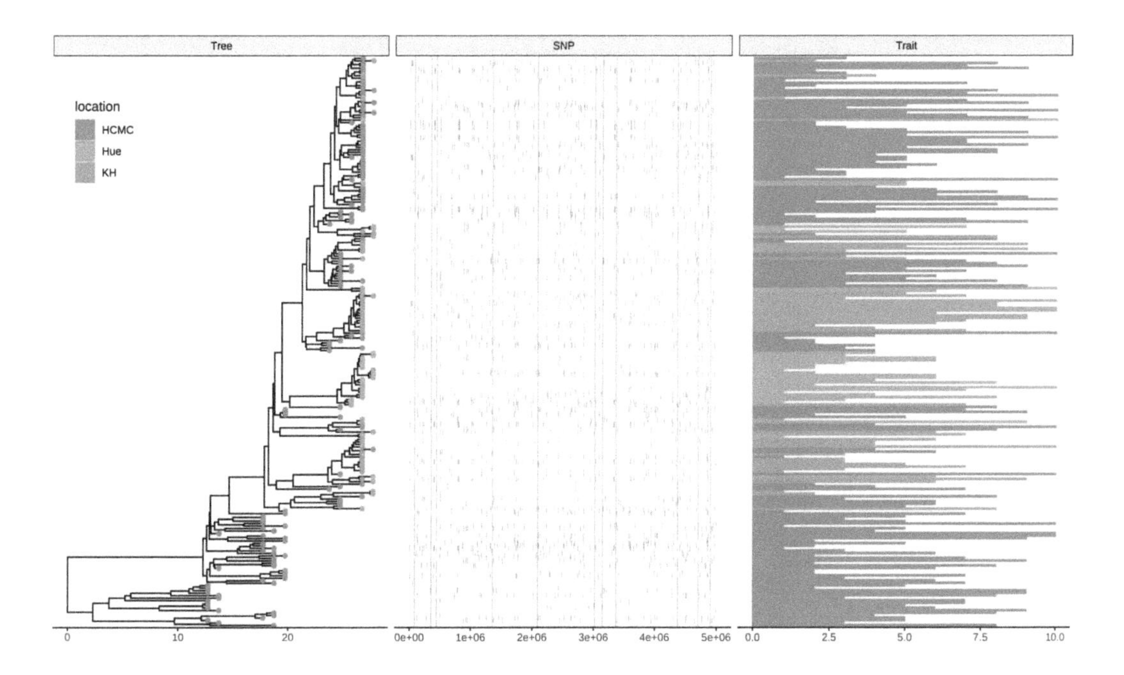

Figure 7.2: **Example of plotting SNP and trait data**. The 'location' information was attached to the tree and used to color tip symbols (Tree panel), and other datasets. SNP and Trait data were visualized as dot chart (SNP panel) and bar chart (Trait panel).

Another difference is that `geom_facet()` only supports rectangular and slanted tree layouts, while `gheatmap()` supports rectangular, slanted, and circular (Figure 7.4) layouts.

In the following example, we visualized a tree of H3 influenza viruses with their associated genotypes (Figure 7.3A).

```
beast_file <- system.file("examples/MCC_FluA_H3.tree", package="ggtree")
beast_tree <- read.beast(beast_file)

genotype_file <- system.file("examples/Genotype.txt", package="ggtree")
genotype <- read.table(genotype_file, sep="\t", stringsAsFactor=F)
colnames(genotype) <- sub("\\.$", "", colnames(genotype))
p <- ggtree(beast_tree, mrsd="2013-01-01") +
    geom_treescale(x=2008, y=1, offset=2) +
    geom_tiplab(size=2)
gheatmap(p, genotype, offset=5, width=0.5, font.size=3,
        colnames_angle=-45, hjust=0) +
    scale_fill_manual(breaks=c("HuH3N2", "pdm", "trig"),
        values=c("steelblue", "firebrick", "darkgreen"), name="genotype")
```

The `width` parameter is to control the width of the heatmap. It supports another parameter `offset` for controlling the distance between the tree and the heatmap, such as allocating space for tip labels.

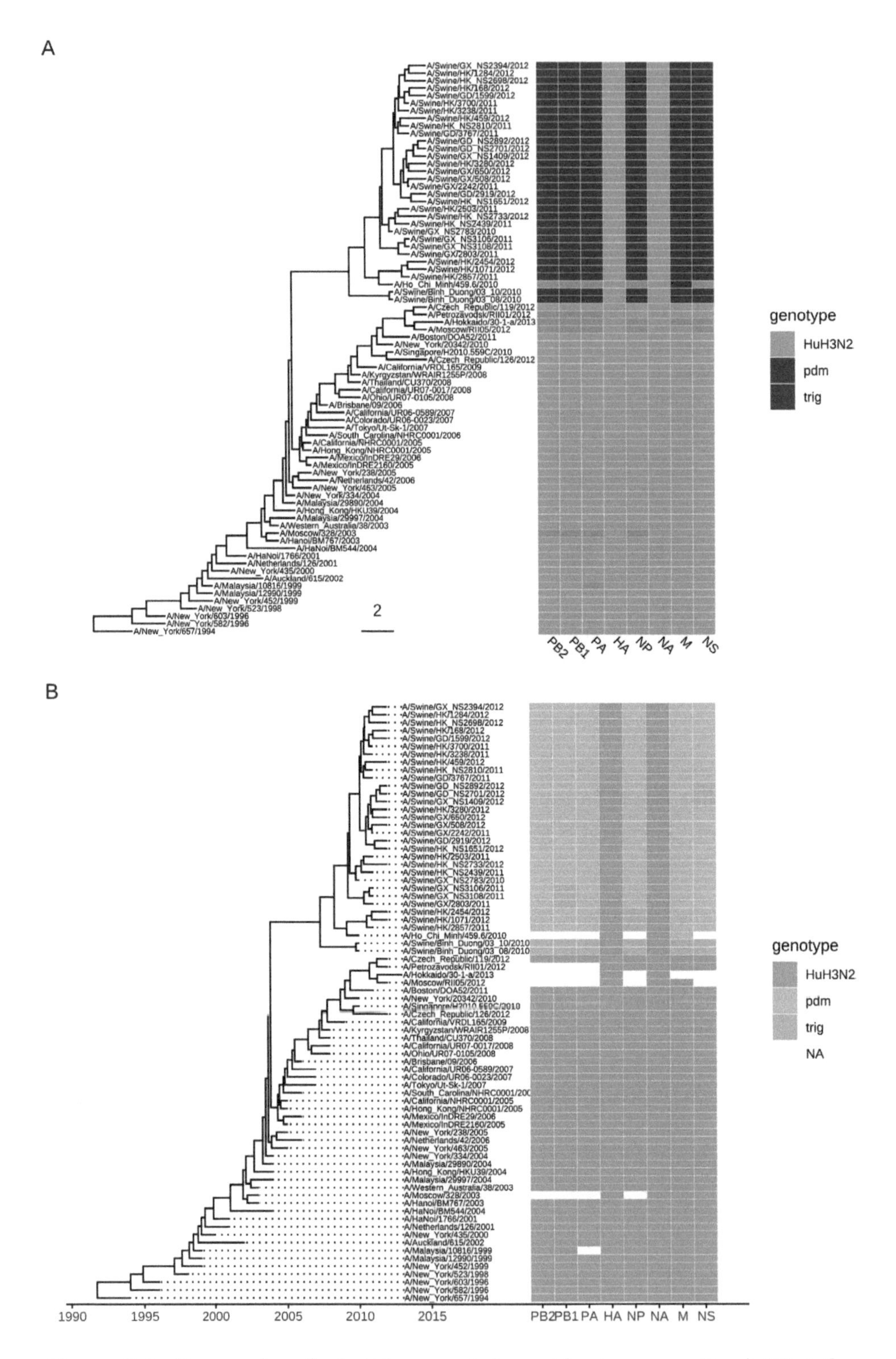

Figure 7.3: **Example of plotting matrix with `gheatmap()`.** A H3 influenza tree with a genotype table visualized as a heatmap (A). Tips were aligned and with a tailored x-axis for divergence times (tree) and genomic segments (heatmap) (B).

For a timescaled tree, as in this example, it's more common to use x-axis by using `theme_tree2`. But with this solution, the heatmap is just another layer and will change the x-axis. To overcome this issue, we implemented `scale_x_ggtree()` to set the x-axis more reasonably (Figure 7.3B).

```
p <- ggtree(beast_tree, mrsd="2013-01-01") +
    geom_tiplab(size=2, align=TRUE, linesize=.5) +
    theme_tree2()
gheatmap(p, genotype, offset=8, width=0.6,
        colnames=FALSE, legend_title="genotype") +
    scale_x_ggtree() +
    scale_y_continuous(expand=c(0, 0.3))
```

7.3.1 Visualize a tree with multiple associated matrices

Of course, we can use multiple `gheatmap()` function calls to align several associated matrices with the tree. However, **ggplot2** doesn't allow us to use multiple `fill` scales[1].

To solve this issue, we can use the **ggnewscale** package to create new `fill` scales. Here is an example of using **ggnewscale** with `gheatmap()`.

```
nwk <- system.file("extdata", "sample.nwk", package="treeio")

tree <- read.tree(nwk)
circ <- ggtree(tree, layout = "circular")

df <- data.frame(first=c("a", "b", "a", "c", "d", "d", "a",
                        "b", "e", "e", "f", "c", "f"),
                 second= c("z", "z", "z", "z", "y", "y",
                        "y", "y", "x", "x", "x", "a", "a"))
rownames(df) <- tree$tip.label

df2 <- as.data.frame(matrix(rnorm(39), ncol=3))
rownames(df2) <- tree$tip.label
colnames(df2) <- LETTERS[1:3]


p1 <- gheatmap(circ, df, offset=.8, width=.2,
               colnames_angle=95, colnames_offset_y = .25) +
    scale_fill_viridis_d(option="D", name="discrete\nvalue")


library(ggnewscale)
```

[1]See also discussion in https://github.com/GuangchuangYu/ggtree/issues/78 and https://groups.google.com/d/msg/bioc-ggtree/VQqbF79NAWU/IjIvpQOBGwAJ

```
p2 <- p1 + new_scale_fill()
gheatmap(p2, df2, offset=15, width=.3,
        colnames_angle=90, colnames_offset_y = .25) +
    scale_fill_viridis_c(option="A", name="continuous\nvalue")
```

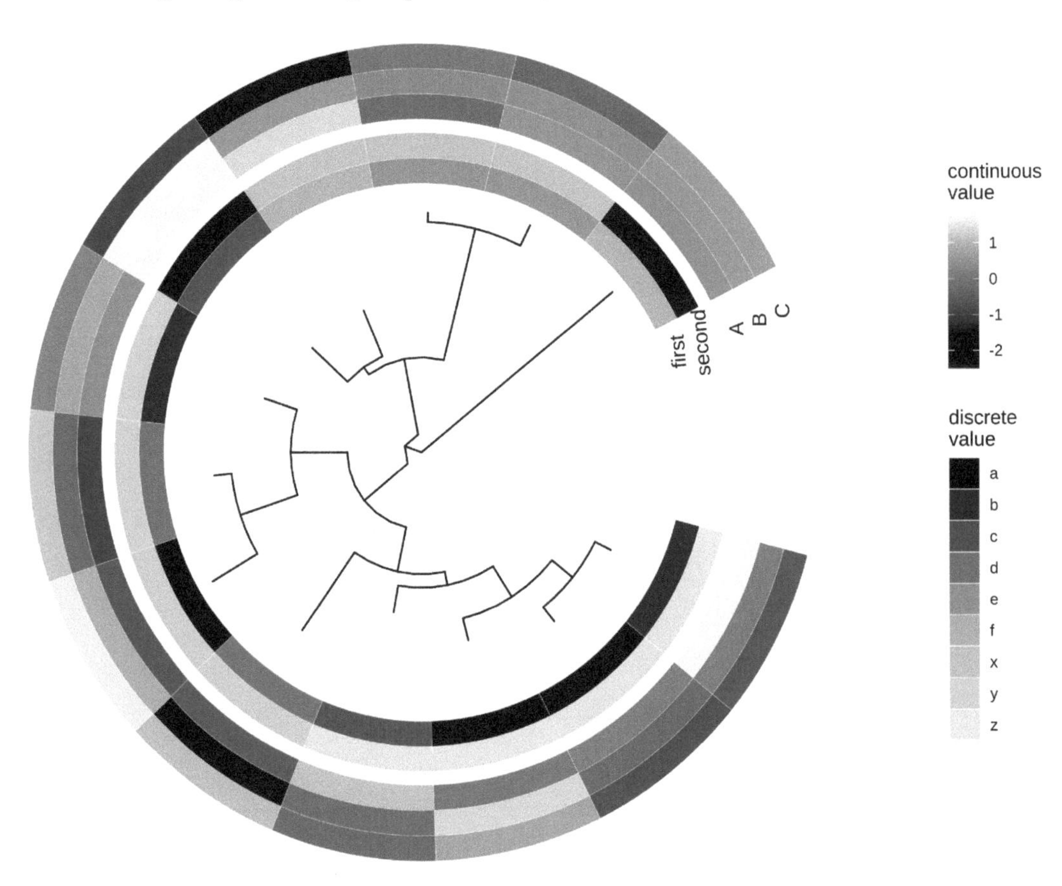

Figure 7.4: **Example of plotting matrix with gheatmap().** A H3 influenza tree with a genotype table visualized as a heatmap (A). Tips were aligned and with a tailored x-axis for divergence times (tree) and genomic segments (heatmap) (B).

7.4 Visualize a Tree with Multiple Sequence Alignments

The `msaplot()` accepts a tree (output of `ggtree()`) and a fasta file, then it can visualize the tree with sequence alignment. We can specify the `width` (relative to the tree) of the alignment and adjust the relative position by `offset`, which is similar to the `gheatmap()` function (Figure 7.5A).

```
library(TDbook)
```

```
# load `tree_seq_nwk` and `AA_sequence` from 'TDbook'
p <- ggtree(tree_seq_nwk) + geom_tiplab(size=3)
msaplot(p, AA_sequence, offset=3, width=2)
```

A specific slice of the alignment can also be displayed by specifying the `window` parameter (Figure 7.5B)..

```
p <- ggtree(tree_seq_nwk, layout='circular') +
    geom_tiplab(offset=4, align=TRUE) + xlim(NA, 12)
msaplot(p, AA_sequence, window=c(120, 200))
```

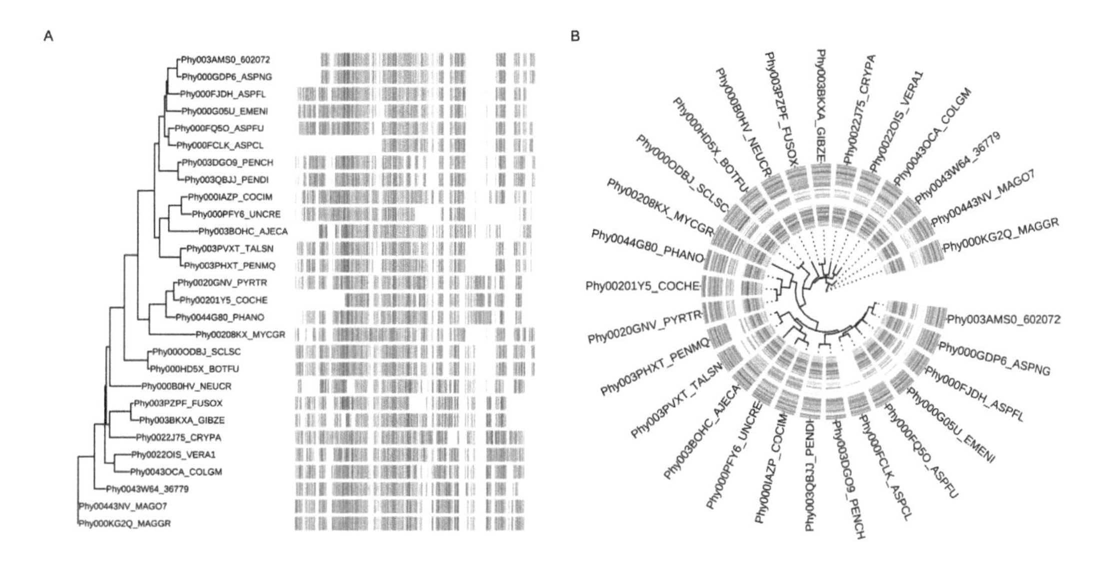

Figure 7.5: **Example of plotting multiple sequence alignments with a tree**. Whole MSA sequences were visualized with a tree in rectangular layout (A). Circular layout with a slice of alignment window (B).

To better support visualizing multiple sequence alignments with a tree and other associated data, we developed the **ggmsa** package with the ability to label the sequences and color the sequences with different color schemes (Yu, 2020). The `ggmsa()` output is compatible with `geom_facet()` and `ggtreeExtra::geom_fruit()` and can be used to visualize a tree, multiple sequence alignments, and different types of associated data to explore their underlying linkages/associations.

7.5 Composite Plots

In addition to aligning graphs to a tree using `geom_facet()` or `ggtreeExtra::geom_fruit()` and special cases using the `gheatmap()` and `msaplot()` functions, users can use **cowplot**, **patchwork**, **gtable**[2] or other packages to create composite

[2] https://github.com/YuLab-SMU/ggtree/issues/313

plots. However, extra efforts need to be done to make sure all the plots are aligned properly. The `ggtree::get_taxa_name()` function is quite useful for users to re-order their data based on the tree structure. To remove this obstacle, we created an R package **aplot** that can re-order the internal data of a `ggplot` object and create composite plots that align properly with a tree.

In the following example, we have a tree with two associated datasets.

```
library(ggplot2)
library(ggtree)

set.seed(2019-10-31)
tr <- rtree(10)

d1 <- data.frame(
    # only some labels match
    label = c(tr$tip.label[sample(5, 5)], "A"),
    value = sample(1:6, 6))

d2 <- data.frame(
    label = rep(tr$tip.label, 5),
    category = rep(LETTERS[1:5], each=10),
    value = rnorm(50, 0, 3))

g <- ggtree(tr) + geom_tiplab(align=TRUE) + hexpand(.01)

p1 <- ggplot(d1, aes(label, value)) + geom_col(aes(fill=label)) +
    geom_text(aes(label=label, y= value+.1)) +
    coord_flip() + theme_tree2() + theme(legend.position='none')

p2 <- ggplot(d2, aes(x=category, y=label)) +
    geom_tile(aes(fill=value)) + scale_fill_viridis_c() +
    theme_minimal() + xlab(NULL) + ylab(NULL)
```

If we align them using **cowplot**, the composite plots are not aligned properly as we anticipated (Figure 7.6A).

```
cowplot::plot_grid(g, p2, p1, ncol=3)
```

Using **aplot**, it will do all the dirty work for us and all the subplots are aligned properly as demonstrated in Figure 7.6B.

```
library(aplot)
p2 %>% insert_left(g) %>% insert_right(p1, width=.5)
```

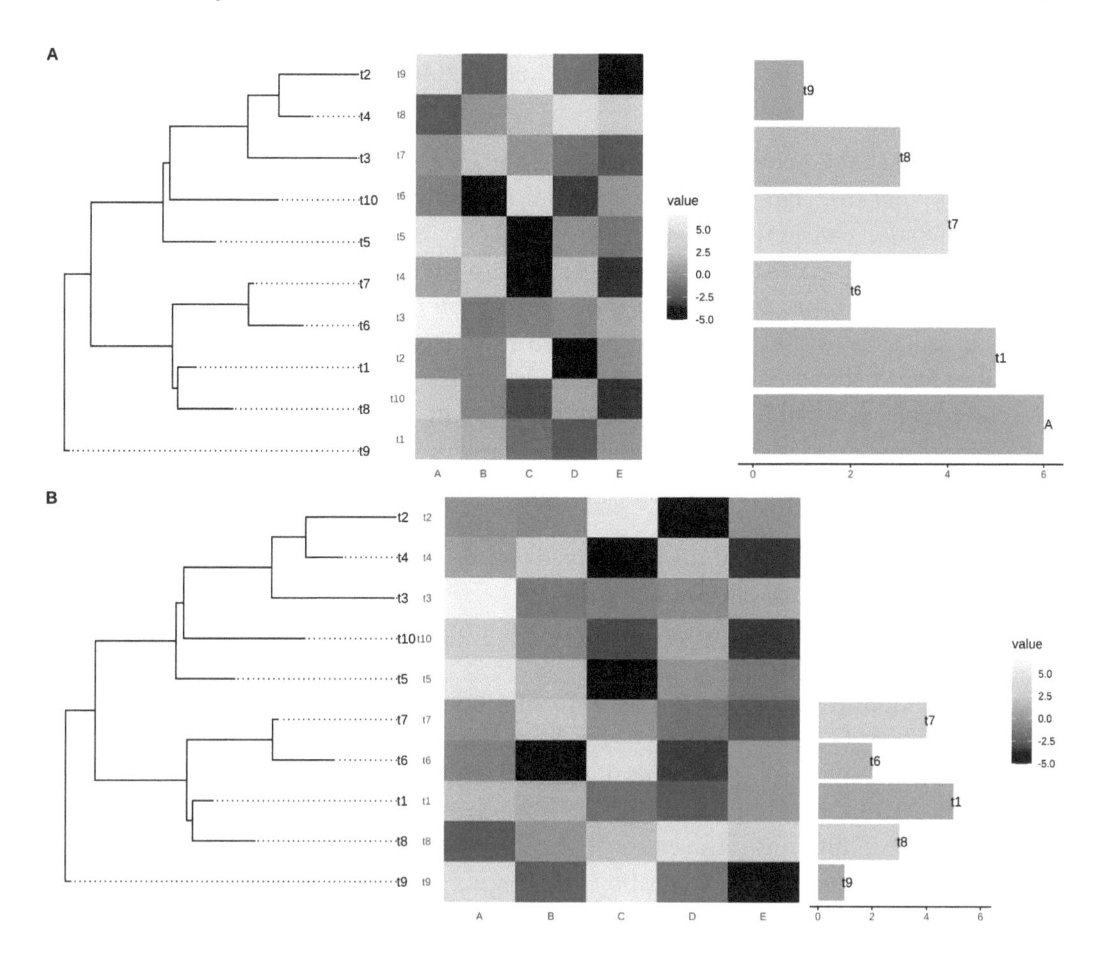

Figure 7.6: **Example of aligning tree with data side-by-side to create composite plot**. **cowplot**‘ just places the subplots together (A), while **aplot** does extra work to make sure that tree-associated subplots are properly ordered according to the tree structure (B). Note: The ‘A’ category in the bar plot that is not matched with the tree was removed.

7.6 Summary

Although there are many software packages that support visualizing phylogenetic trees, plotting a tree with data is often missing or with only limited support. Some of the packages define `S4` classes to store phylogenetic tree with domain-specific data, such as **OutbreakTools** (Jombart et al., 2014) defined `obkData` for storing tree with epidemiology data and **phyloseq** (McMurdie & Holmes, 2013) defines `phyloseq` for storing tree with microbiome data. These packages are capable of presenting some of the data stored in the object on the tree. However, not all the associated data are supported. For example, species abundance stored in the `phyloseq` object is not supported to be visualized using the **phyloseq** package. These packages did not provide any utilities to integrate external data for tree visualization. None of these packages support visualizing external data and aligning the plot to a tree based on the tree structure.

The **ggtree** package provides two general solutions for integrating data. Method 1, the `%<+%` operator, can integrate external and internal node data and map the data as a visual characteristic to visualize the tree and other datasets used in `geom_facet()` or `ggtreeExtra::geom_fruit()`. Method 2, the `geom_facet` layer or `ggtreeExtra::geom_fruit()`, has no restriction of input data as long as there is a `geom` function available to plot the data (*e.g.*, species abundance displayed by `geom_density_ridges` as demonstrated in Figure 9.4). Users are free to combine different panels and combine different `geom` layers in the same panel (Figure 13.1).

The **ggtree** package has many unique features that cannot be found in other implementations (Yu et al., 2018):

1. Integrating node/edge data to the tree can be mapped to visual characteristics of the tree or other datasets (Figure 7.1).
2. Capable of parsing expressions (math symbols or text formatting), emoji, and image files (Chapter 8).
3. No pre-definition of input data types or how the data should be plotted in `geom_facet()` (Table 1).
4. Combining different `geom` functions to visualize associated data is supported (Figure 13.1).
5. Visualizing different datasets on the same panel is supported.
6. Data integrated by `%<+%` can be used in `geom_facet()` layer.
7. Able to add further annotations to specific layers.
8. Modular design by separating tree visualization, data integration (Method 1), and graph alignment (Method 2).

Modular design is a unique feature for **ggtree** to stand out from other packages. The tree can be visualized with data stored in the tree object or external data linked by the `%<+%` operator, and fully annotated with multiple layers of annotations (Figures 7.1 and 13.1), before passing it to `geom_facet()` layer. The `geom_facet()` layer can be called progressively to add multiple panels or multiple layers on the same panels (Figure 13.1). This creates the possibility of plotting a full annotated tree with complex data panels that contain multiple graphic layers.

The **ggtree** package fits the `R` ecosystem and extends the abilities to integrate and present data with trees to existing phylogenetic packages. As demonstrated in Figure 9.4, we can plot species abundance distributions with the `phyloseq` object. This cannot be easily done without **ggtree**. With **ggtree**, we are able to attach additional data to tree objects using the `%<+%` operator and align graphs to a tree using the `geom_facet()` layer. Integrating **ggtree** into existing workflows will extend the abilities and broaden the applications to present phylogeny-associated data, especially for comparative studies.

Chapter 8

Annotating Tree with Silhouette Images and Sub-plots

8.1 Annotating Tree with Images

We usually use text to label taxa, *i.e.* displaying taxa names. If the text is the image file name (either local or remote), **ggtree** can read the image and display the actual image as the label of the taxa (Figure 8.1). The `geom_tiplab()` and `geom_nodelab()` are capable to render silhouette images with supports from in-house developed package, **ggimage**.

Online tools such as **iTOL** (Letunic & Bork, 2007) and **EvolView** (He et al., 2016) support displaying subplots on a phylogenetic tree. However, only bar and pie charts are supported by these tools. Users may want to visualize node-associated data with other visualization methods, such as violin plot (Grubaugh et al., 2017), venn diagram (Lott et al., 2015), sequence logo, *etc.*, and display them on the tree. In **ggtree**, all kinds of subplots are supported, as we can export all subplots to image files and use them to label corresponding nodes on the tree.

```
library(ggimage)
library(ggtree)

nwk <- paste0("((((bufonidae, dendrobatidae), ceratophryidae),",
          "(centrolenidae, leptodactylidae)), hylidae);")

imgdir <- system.file("extdata/frogs", package = "TDbook")

x = read.tree(text = nwk)
ggtree(x) + xlim(NA, 7) + ylim(NA, 6.2) +
    geom_tiplab(aes(image=paste0(imgdir, '/', label, '.jpg')),
                geom="image", offset=2, align=2, size=.2)  +
```

```
geom_tiplab(geom='label', offset=1, hjust=.5) +
geom_image(x=.8, y=5.5, image=paste0(imgdir, "/frog.jpg"), size=.2)
```

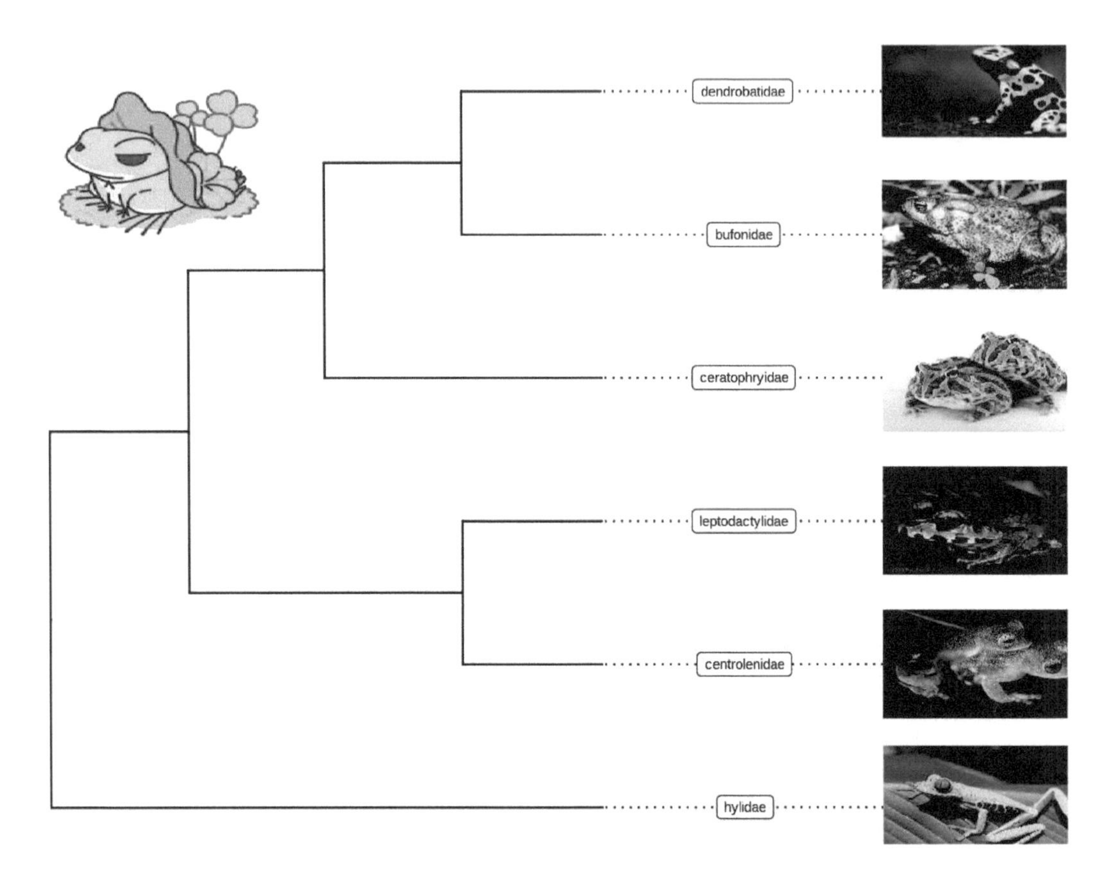

Figure 8.1: **Labelling taxa with images.** Users need to specify `geom = "image"` and map the image file names onto the `image` aesthetics.

8.2 Annotating Tree with Phylopic

Phylopic contains more than 3200 silhouettes and covers almost all life forms. The **ggtree** package supports using phylopic[1] to annotate the tree by setting `geom="phylopic"` and mapping phylopic UID to the `image` aesthetics. The **ggimage** package supports querying phylopic UID from the scientific name, which is very handy for annotating tree with phylopic. In the following example, tip labels were used to query phylopic UID, and phylopic images were used to label the tree as another layer of tip labels. Most importantly, we can color or resize the images using numerical/categorical variables, and here the values of body mass were used to encode the color of the images (Figure 8.2).

[1]https://twitter.com/guangchuangyu/status/593443854541434882

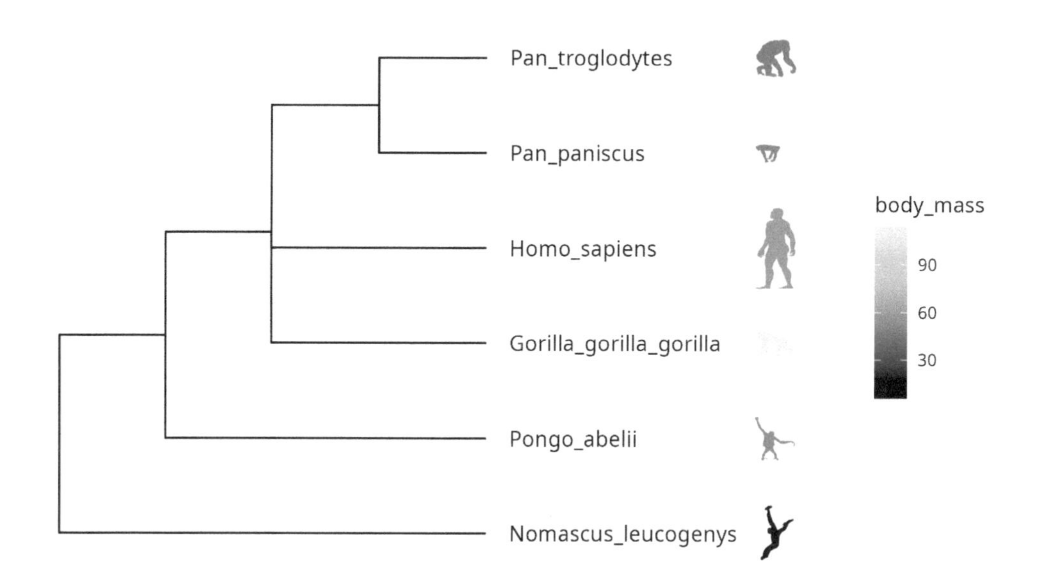

Figure 8.2: **Labelling taxa with phylopic images.** The **ggtree** will automatically download phylopic figures by querying provided UID. The figures can be colored using numerical or categorical values.

```
library(ggtree)
newick <- paste0("((Pongo_abelii,(Gorilla_gorilla_gorilla,(Pan_paniscus,",
          "Pan_troglodytes)Pan,Homo_sapiens)Homininae)Hominidae,",
          "Nomascus_leucogenys)Hominoidea;")

tree <- read.tree(text=newick)

d <- ggimage::phylopic_uid(tree$tip.label)
d$body_mass <- c(52, 114, 47, 45, 58, 6)

p <- ggtree(tree) %<+% d +
  geom_tiplab(aes(image=uid, colour=body_mass),geom="phylopic", offset=2.5)+
  geom_tiplab(aes(label=label), offset = .2) + xlim(NA, 7) +
  scale_color_viridis_c()
```

8.3 Annotating Tree with Sub-plots

The **ggtree** package provides a layer, `geom_inset()`, for adding subplots to a phylogenetic tree. The input is a named list of `ggplot` graphic objects (can be any kind of chart). These objects should be named by node numbers. Users can also use **ggplotify** to convert plots generated by other functions (even implemented by base graphics) to `ggplot` objects, which can then be used in the `geom_inset()` layer. To facilitate adding bar and pie charts (*e.g.*, summarized stats of results from ancestral reconstruction) to the phylogenetic tree, **ggtree** provides the `nodepie()` and `nodebar()` functions to create a list of pie or bar charts.

8.3.1 Annotate with bar charts

This example uses `ape::ace()` function to estimate ancestral character states. The likelihoods of the stats were visualized as stacked bar charts which were overlayed onto internal nodes of the tree using the `geom_inset()` layer (Figure 8.3A).

```
library(phytools)
data(anoletree)
x <- getStates(anoletree,"tips")
tree <- anoletree

cols <- setNames(palette()[1:length(unique(x))],sort(unique(x)))
fitER <- ape::ace(x,tree,model="ER",type="discrete")
ancstats <- as.data.frame(fitER$lik.anc)
ancstats$node <- 1:tree$Nnode+Ntip(tree)

## cols parameter indicate which columns store stats
bars <- nodebar(ancstats, cols=1:6)
bars <- lapply(bars, function(g) g+scale_fill_manual(values = cols))

tree2 <- full_join(tree, data.frame(label = names(x), stat = x ),
                   by = 'label')
p <- ggtree(tree2) + geom_tiplab() +
    geom_tippoint(aes(color = stat)) +
    scale_color_manual(values = cols) +
    theme(legend.position = "right") +
    xlim(NA, 8)
p1 <- p + geom_inset(bars, width = .08, height = .05, x = "branch")
```

The `x` position can be one of `node' orbranch' and can be adjusted by the parameters, `hjust` and `vjust`, for horizontal and vertical adjustment, respectively. The `width` and `height` parameters restrict the size of the inset plots.

8.3.2 Annotate with pie charts

Similarly, users can use the `nodepie()` function to generate a list of pie charts and place these charts to annotate corresponding nodes (Figure 8.3B). Both `nodebar()` and `nodepie()` accept a parameter of `alpha` to allow transparency.

```
pies <- nodepie(ancstats, cols = 1:6)
pies <- lapply(pies, function(g) g+scale_fill_manual(values = cols))
p2 <- p + geom_inset(pies, width = .1, height = .1)

plot_list(p1, p2, guides='collect', tag_levels='A')
```

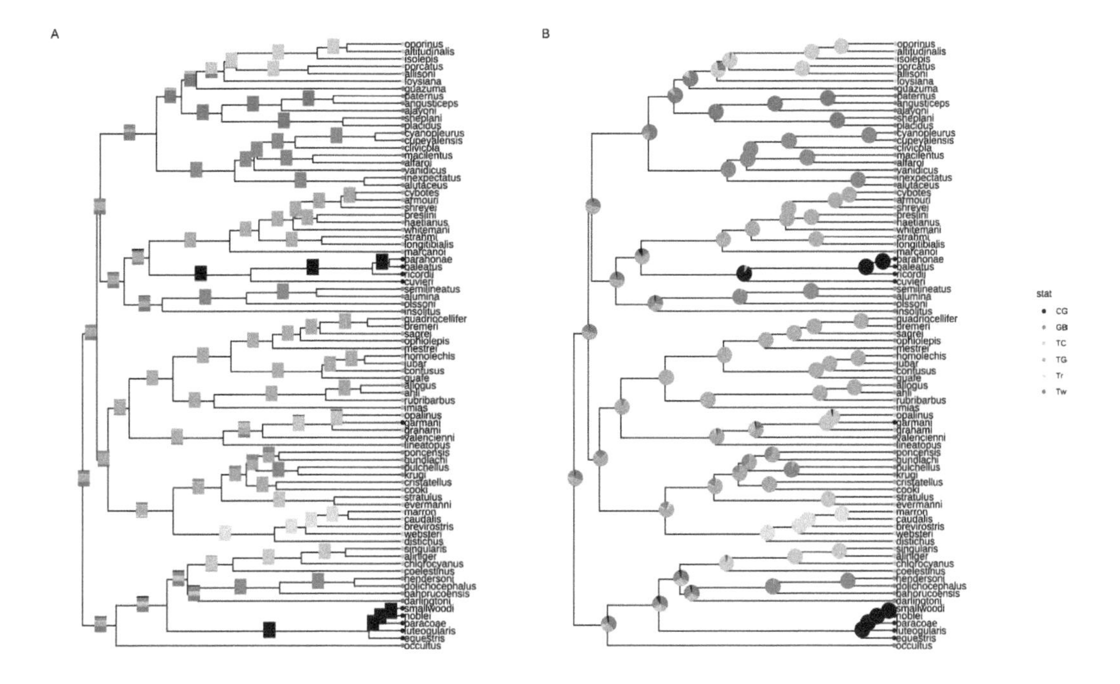

Figure 8.3: **Annotate internal nodes with bar or pie charts.** Using bar charts (A) or pie charts (B) to display summary statistics of internal nodes.

8.3.3 Annotate with mixed types of charts

The `geom_inset()` layer accepts a list of `ggplot' graphic objects and these input objects are not restricted to pie or bar charts. They can be any kind of charts or hybrid of these charts. The`geom_inset()‘ is not only useful to display ancestral stats, but also applicable to visualize different types of data that are associated with selected nodes in the tree. Here, we use a mixture of pie and bar charts to annotate the tree as an example (Figure 8.4).

```
pies_and_bars <- pies
i <- sample(length(pies), 20)
pies_and_bars[i] <- bars[i]
p + geom_inset(pies_and_bars, width=.08, height=.05)
```

8.4 Have Fun with Phylomoji

Phylomoji is a phylogenetic tree of emoji. It is fun[2] and very useful for education of the evolution concept. The **ggtree** supports producing phylomoji since 2015[3]. Here, we will use **ggtree** to recreate the following phylomoji figure[4] (Figure 8.5):

[2]https://twitter.com/hashtag/phylomoji?src=hash

[3]https://twitter.com/guangchuangyu/status/662095056610811904 and https://twitter.com/guangchuangyu/status/667337429704011777

[4]https://twitter.com/OxyMLZ/status/1055586178651451392

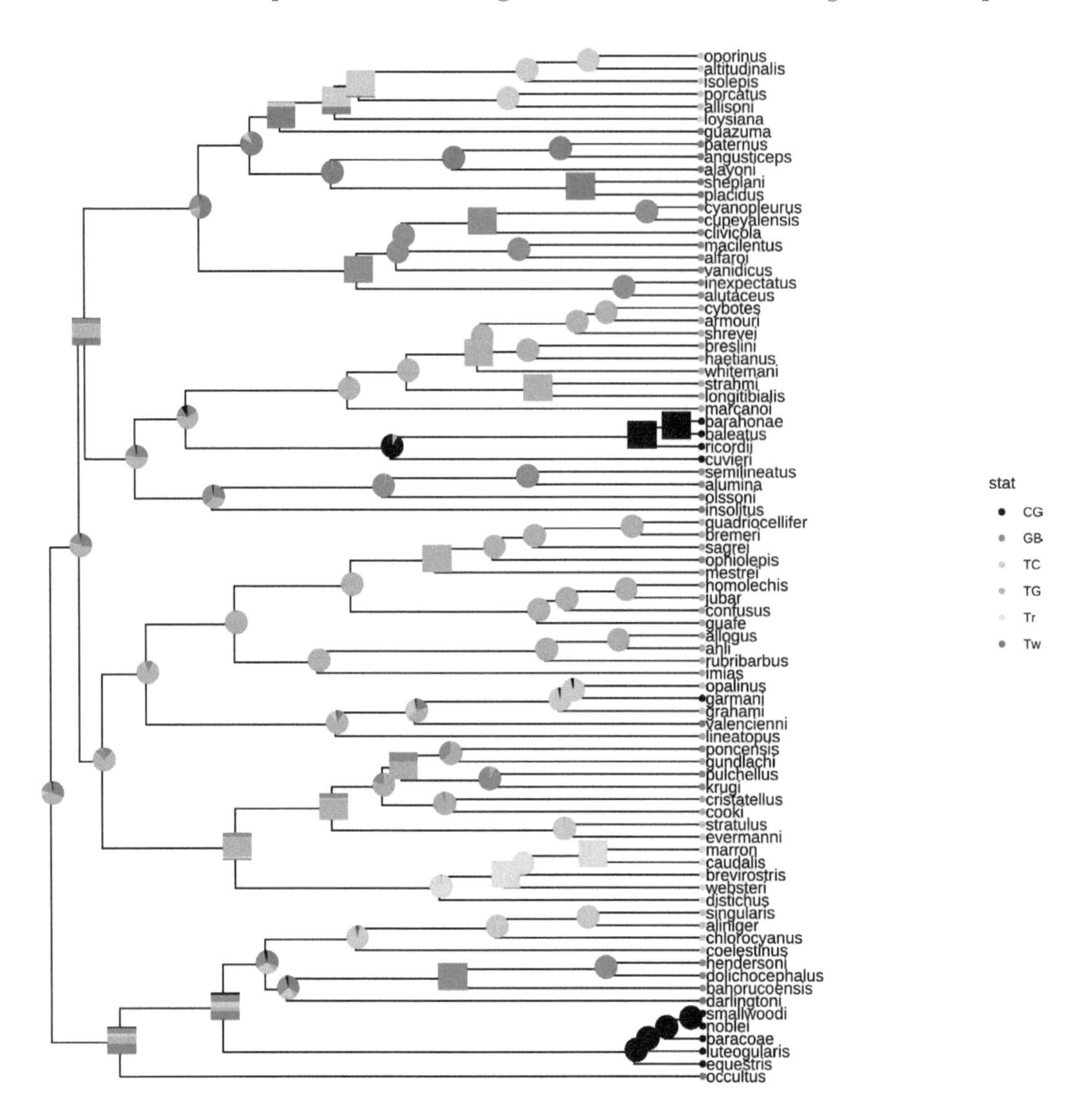

Figure 8.4: **Annotate internal nodes with different types of subplots.**

```
library(ggplot2)
library(ggtree)

tt = '((snail,mushroom),(((sunflower,evergreen_tree),leaves),green_salad));'
tree = read.tree(text = tt)
d <- data.frame(label = c('snail','mushroom', 'sunflower',
                          'evergreen_tree','leaves', 'green_salad'),
                group = c('animal', 'fungi', 'flowering plant',
                          'conifers', 'ferns', 'mosses'))

p <- ggtree(tree, linetype = "dashed", size=1, color='firebrick') %<+% d +
  xlim(0, 4.5) + ylim(0.5, 6.5) +
```

```
  geom_tiplab(parse="emoji", size=15, vjust=.25) +
  geom_tiplab(aes(label = group), geom="label", x=3.5, hjust=1)
```

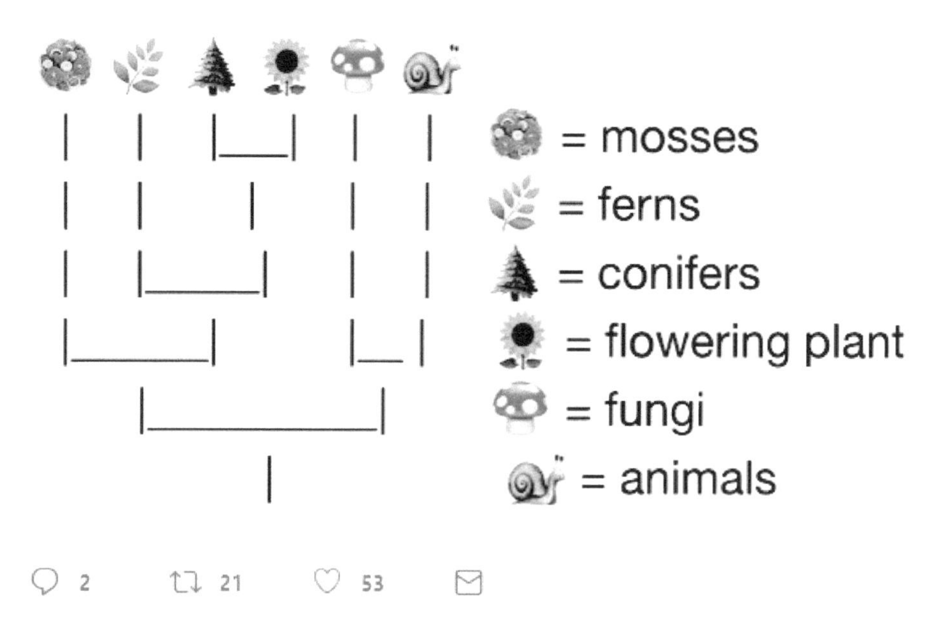

Note that the output may depend on what emoji fonts are installed in your system[5].

With **ggtree**, it is easy to generate phylomoji. The emoji is treated as text, like 'abc'. We can use emojis to label taxa, clade, color and rotate emoji with any given color and angle. This functionality is internally supported by the **emojifont** package.

8.4.1 Emoji in circular/fan layout tree

It also works with circular and fan layouts as demonstrated in Figure 8.6.

```
p <- ggtree(tree, layout = "circular", size=1) +
  geom_tiplab(parse="emoji", size=10, vjust=.25)
print(p)

## fan layout
p2 <- open_tree(p, angle=200)
print(p2)

p2 %>% rotate_tree(-90)
```

Another example using **ggtree** and **emojifont** to produce phylogeny of plant emojis can be found in a scientific article (Escudero & Wendel, 2020).

[5]Google Noto emoji fonts was installed in my system

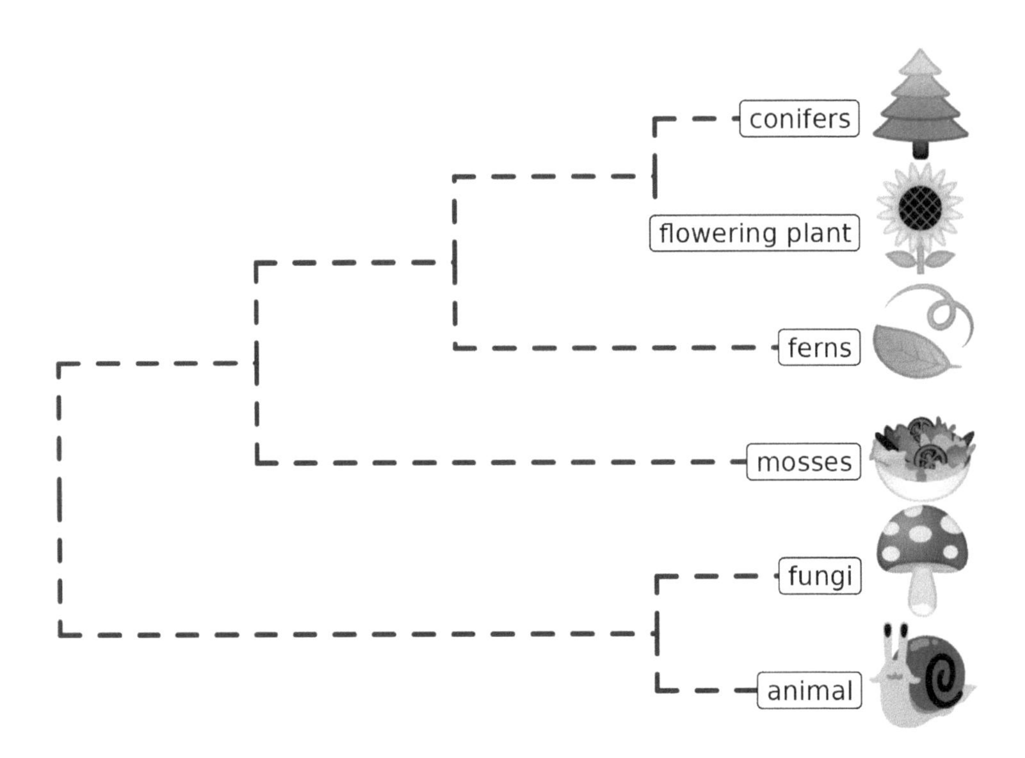

Figure 8.5: **Parsing label as emoji.** Text (*e.g.*, node or tip labels) can be parsed as emoji.

8.4.2 Emoji to label clades

Parsing clade labels as emojis is also supported in the `geom_cladelab()` layer. For example, in a phylogenetic tree of influenza viruses, we can use emojis to label clades to represent host species similar to Figure 8.7.

```
set.seed(123)
tr <- rtree(30)

dat <- data.frame(
           node = c(41, 53, 48),
           name = c("chicken", "duck", "family")
       )

p <- ggtree(tr) +
     xlim(NA, 5.2) +
     geom_cladelab(
         data = dat,
```

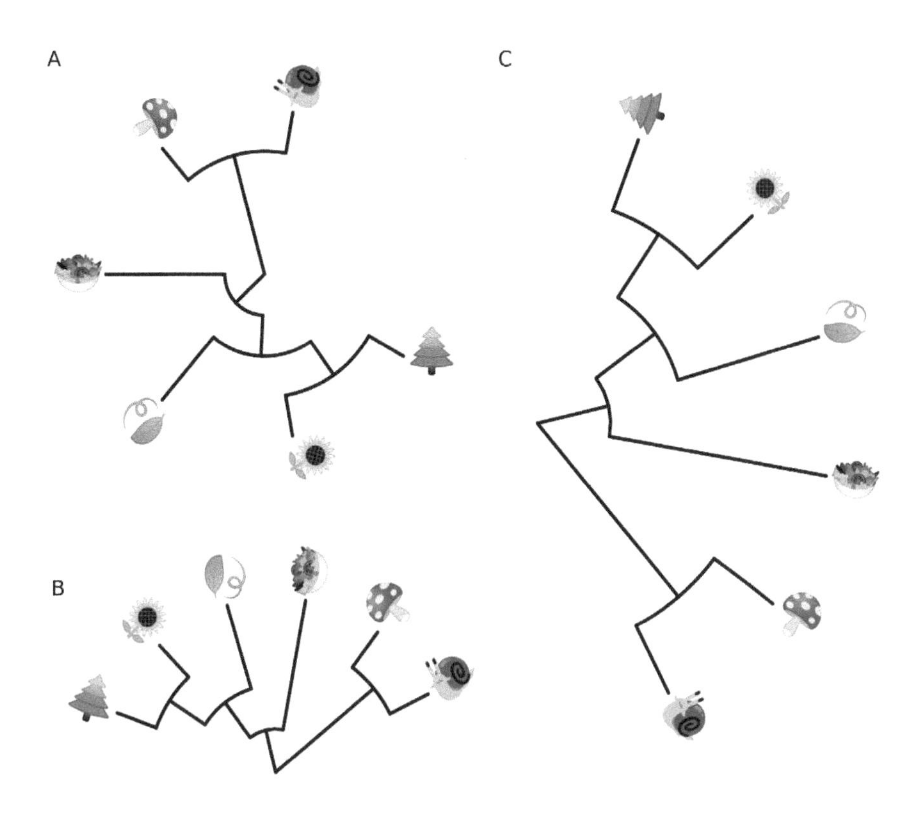

Figure 8.6: **Phylomoji in cirular and fan layouts.**

```
    mapping = aes(
        node = node,
        label = name,
        color = name
    ),
    parse = "emoji",
    fontsize = 12,
    align = TRUE,
    show.legend = FALSE
) +
scale_color_manual(
    values = c(
        chicken="firebrick",
        duck="steelblue",
        family = "darkkhaki"
    )
)
```

```
p
```

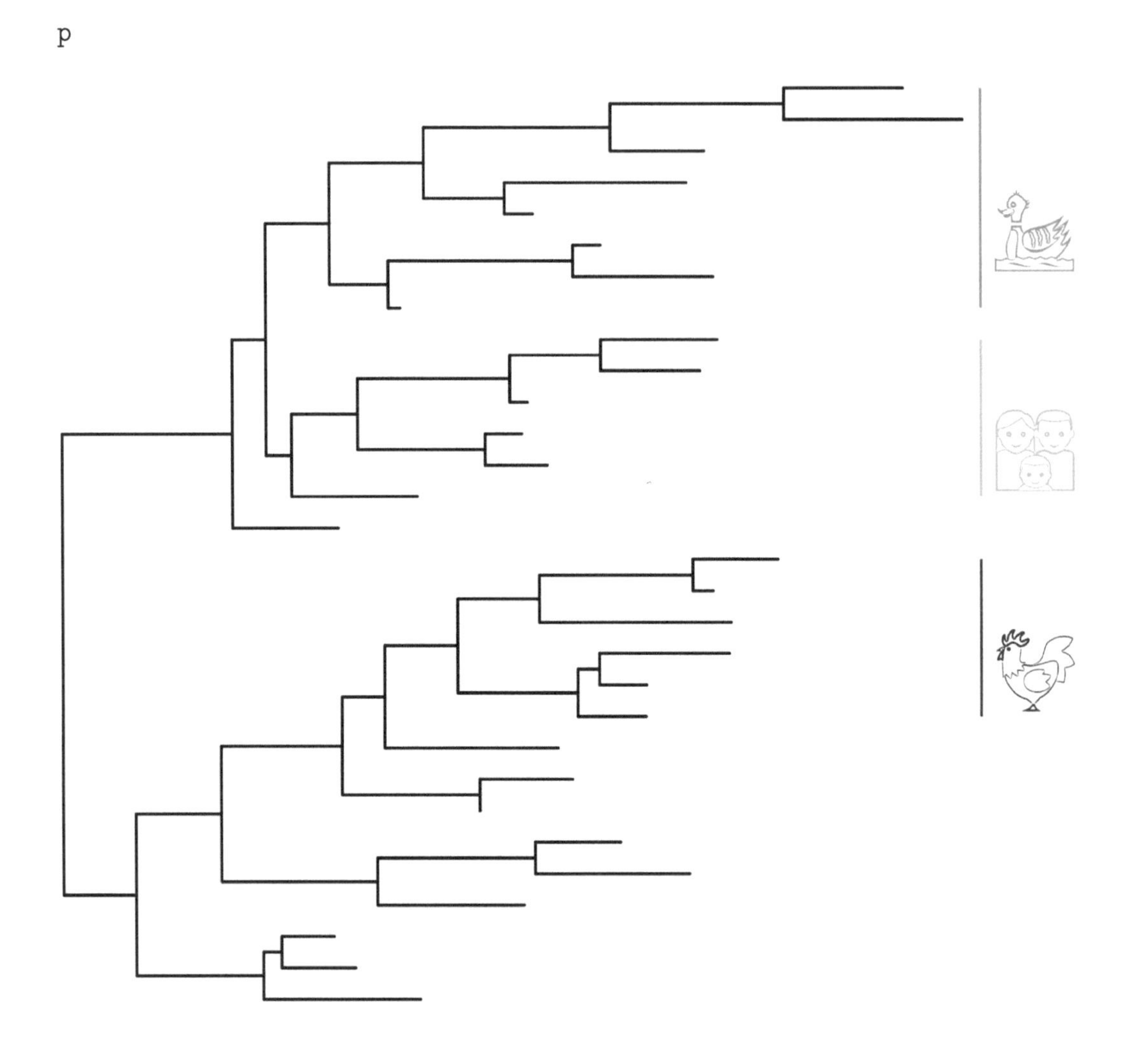

Figure 8.7: **Emoji to label clades.**

8.4.3 Apple Color Emoji

Although R's graphical devices don't support AppleColorEmoji font on MacOS, it's still possible to use it. We can export the plot to a svg file and render it in Safari (Figure 8.8).

```
library(ggtree)
tree_text <- paste0("(((((cow, (whale, dolphin)), (pig2, boar)),",
                    "camel), fish), seedling);")
x <- read.tree(text=tree_text)
library(ggimage)
p <-  ggtree(x, size=2) + geom_tiplab(size=20, parse='emoji') +
    xlim(NA, 7) + ylim(NA, 8.5)
```

```
svglite::svglite("emoji.svg", width = 10, height = 7)
print(p)
dev.off()

# or use `grid.export()`
# ps = gridSVG::grid.export("emoji.svg", addClass=T)
```

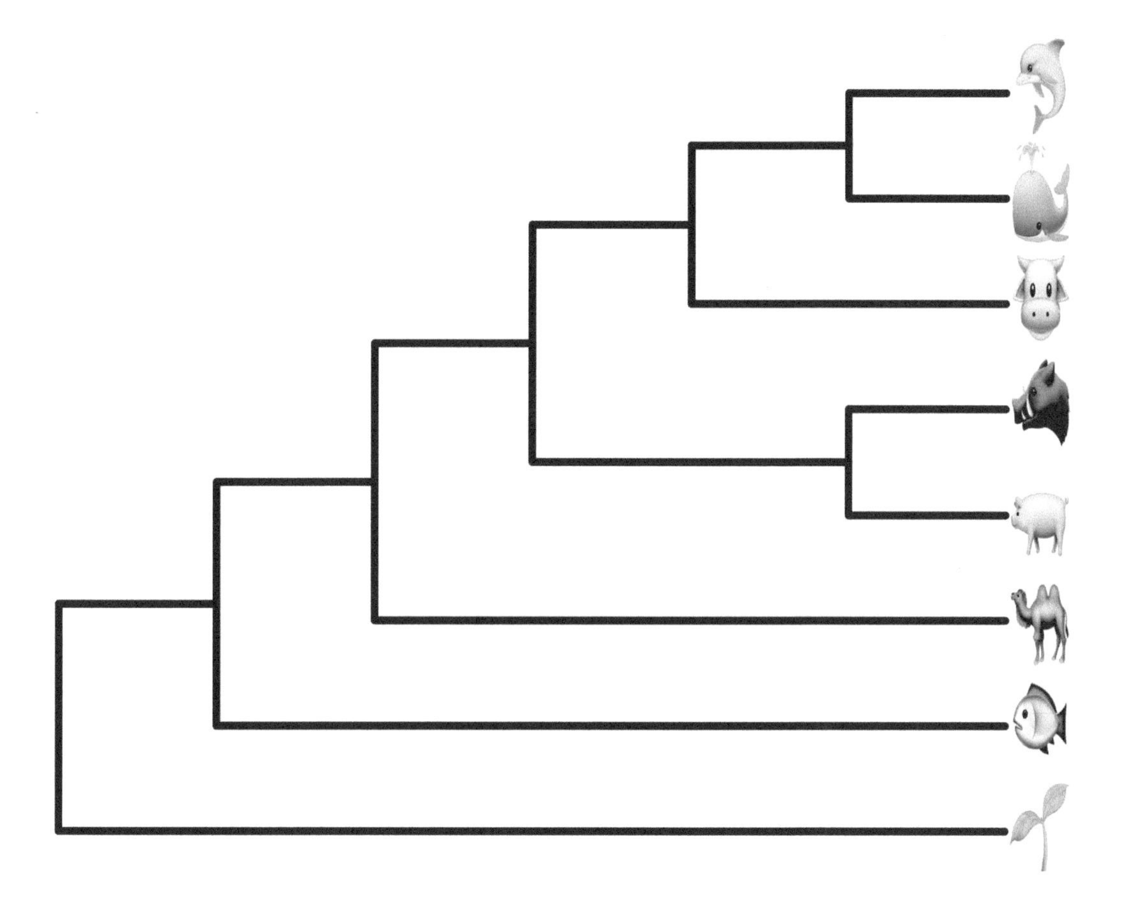

Figure 8.8: **Use Apple Color Emoji in ggtree.** The tip labels were parsed as emojis using `AppleColorEmoji` font in Safari.

8.4.4 Phylomoji in ASCII art

Producing phylomoji as an ASCII art is also possible. Users can refer to Appendix D for details.

8.5 Summary

The **ggtree** supports parsing labels, including tip labels, internal node labels, and clade labels, as images, math expression, and emoji, in case the labels can be parsed as image file names, `plotmath` expression, or emoji names, respectively. It can be fun, but it's also very useful for scientific research. The use of images on phylogenetic trees can help to present species-related characteristics, including morphological, anatomical, and even macromolecular structures. Moreover, **ggtree** supports summarizing statistical inferences (*e.g.*, biogeographic range reconstruction and posterior distribution) or associated data of the nodes as subplots to be displayed on a phylogenetic tree.

Chapter 9

ggtree for Other Tree-like Objects

9.1 ggtree for Phylogenetic Tree Objects

The **treeio** packages (Wang et al., 2020) allow parsing evolutionary inferences from several software outputs and linking external data to the tree structure. It serves as an infrastructure to bring evolutionary data to the R community. The **ggtree** package (Yu et al., 2017) works seamlessly with **treeio** to visualize tree-associated data to annotate the tree. The **ggtree** package is a general tool for tree visualization and annotation and it fits the ecosystem of R packages. Most of the S3/S4 tree objects defined by other R packages are also supported by **ggtree**, including `phylo` (session 4.2), `multiPhylo` (session 4.4), `phylo4`, `phylo4d`, `phyloseq`, and `obkData`. With **ggtree**, we are able to generate more complex tree graphs which is not possible or easy to do with other packages. For example, the visualization of the `phyloseq` object in Figure 9.4 is not supported by the **phyloseq** package. The **ggtree** package also extends the possibility of linking external data to these tree objects (Yu et al., 2018).

9.1.1 The phylo4 and phylo4d objects

The `phylo4` and `phylo4d` are defined in the **phylobase** package. The `phylo4` object is an S4 version of `phylo`, while `phylo4d` extends `phylo4` with a data frame that contains trait data. The **phylobase** package provides a `plot()` method, which is internally called the `treePlot()` function, to display the tree with the data. However, there are some restrictions of the `plot()` method, it can only plot numeric values for tree-associated data as bubbles and cannot generate figure legend. `Phylobase` doesn't implement a visualization method to display categorical values. Using associated data as visual characteristics such as color, size, and shape, is also not supported. Although it is possible to color the tree using associated data, it requires users to extract the data and map them to the color vector manually followed by passing the color vector to the `plot` method. This is tedious and error-prone since the order of the color vector needs to be consistent with the edge list stored in the object.

The **ggtree** package supports `phylo4d` object and all the associated data stored in the `phylo4d` object can be used directly to annotate the tree (Figure 9.1).

```
library(phylobase)
data(geospiza_raw)
g1 <- as(geospiza_raw$tree, "phylo4")
g2 <- phylo4d(g1, geospiza_raw$data, missing.data="warn")

d1 <- data.frame(x = seq(1.1, 2, length.out = 5),
                 lab = names(geospiza_raw$data))

p1 <- ggtree(g2) + geom_tippoint(aes(size = wingL), x = d1$x[1], shape = 1)+
    geom_tippoint(aes(size = tarsusL), x = d1$x[2], shape = 1) +
    geom_tippoint(aes(size = culmenL), x = d1$x[3], shape = 1) +
    geom_tippoint(aes(size = beakD),   x = d1$x[4], shape = 1) +
    geom_tippoint(aes(size = gonysW),  x = d1$x[5], shape = 1) +
    scale_size_continuous(range = c(3,12), name="") +
    geom_text(aes(x = x, y = 0, label = lab), data = d1, angle = 45) +
    geom_tiplab(offset = 1.3) + xlim(0, 3) +
    theme(legend.position = c(.1, .75)) + vexpand(.05, -1)

## users can use `as.treedata(g2)` to convert `g2` to a `treedata` object
## and use `get_tree_data()` function to extract the associated data

p2 <- gheatmap(ggtree(g1), data=geospiza_raw$data, colnames_angle=45) +
  geom_tiplab(offset=1) + hexpand(.2) + vexpand(.05, -1) +
  theme(legend.position = c(.1, .75))

aplot::plot_list(p1, p2, ncol=2, tag_levels='A')
```

9.1.2 Thc phylog object

The `phylog` is defined in the **ade4** package. The package is designed for analyzing ecological data and provides `newick2phylog()`, `hclust2phylog()`, and `taxo2phylog()` functions to create phylogeny from Newick string, hierarchical clustering result, or a taxonomy (see also the **MicrobiotaProcess** package described in Chapter 11). The `phylog` object is also supported by **ggtree** as demonstrated in Figure 9.2.

```
library(ade4)
data(taxo.eg)
tax <- as.taxo(taxo.eg[[1]])
names(tax) <- c("genus", "family", "order")
print(tax)

##       genus family order
## esp3     g1   fam1  ORD1
```

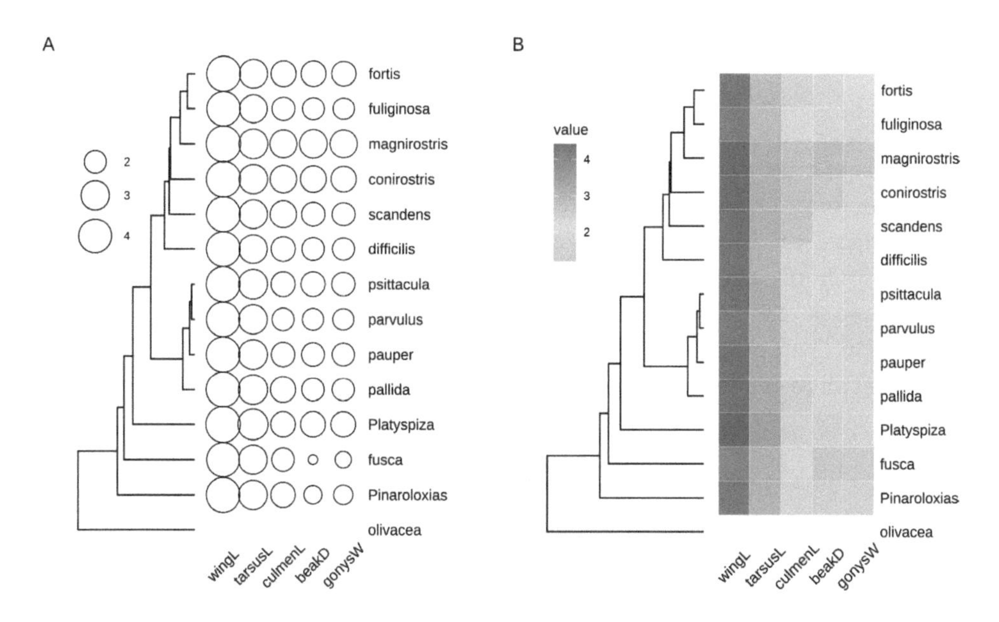

Figure 9.1: **Visualizing `phylo4d` data using ggtree.** Reproduce the output of the `plot()` method provided in the **phylobase** package (A). Visualize the trait data as a heatmap which is not supported in the **phylobase** package (B).

```
## esp1     g1   fam1  ORD1
## esp2     g1   fam1  ORD1
## esp4     g1   fam1  ORD1
## esp5     g1   fam1  ORD1
## esp6     g1   fam1  ORD1
## esp7     g1   fam1  ORD1
## esp8     g2   fam2  ORD2
## esp9     g3   fam2  ORD2
## esp10    g4   fam3  ORD2
## esp11    g5   fam3  ORD2
## esp12    g6   fam4  ORD2
## esp13    g7   fam4  ORD2
## esp14    g8   fam5  ORD2
## esp15    g8   fam5  ORD2
```

```
tax.phy <- taxo2phylog(as.taxo(taxo.eg[[1]]))
print(tax.phy)
```

```
## Phylogenetic tree with 15 leaves and 16 nodes
## $class: phylog
## $call: taxo2phylog(taxo = as.taxo(taxo.eg[[1]]))
## $tre: ((((esp3,esp1,esp2,esp4,e...15)I1g8)I2fam5)I3ORD2)Root;
##
##         class   length
## $leaves numeric 15
```

```
## $nodes  numeric 16
## $parts  list    16
## $paths  list    31
## $droot  numeric 31
##         content
## $leaves length of the first preceeding adjacent edge
## $nodes  length of the first preceeding adjacent edge
## $parts  subsets of descendant nodes
## $paths  path from root to node or leave
## $droot  distance to root
```

```
ggtree(tax.phy) + geom_tiplab() +
  geom_nodelab(geom='label') + hexpand(.05)
```

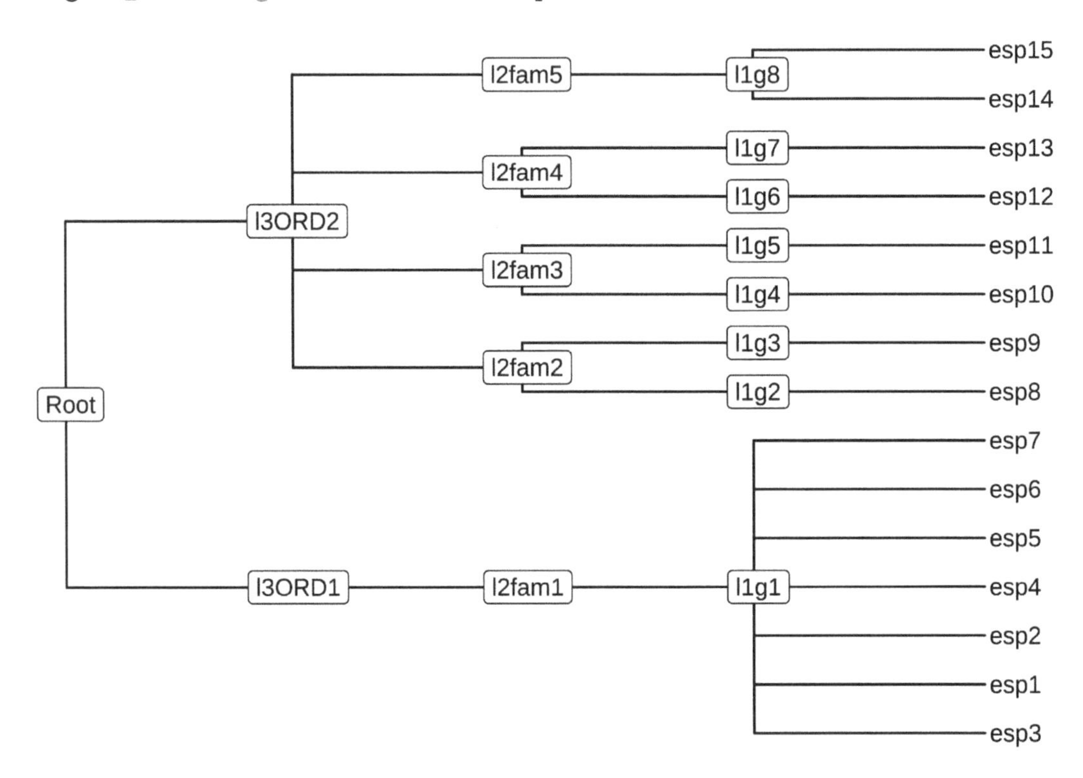

Figure 9.2: **Visualizing a `phylog` tree object.**

9.1.3 The phyloseq object

The `phyloseq` class defined in the **phyloseq** package was designed for storing microbiome data, including a phylogenetic tree, associated sample data, and taxonomy assignment. It can import data from popular pipelines, such as **QIIME** (Kuczynski et al., 2011), **mothur** (Schloss et al., 2009), **dada2** (Callahan et al., 2016) and **PyroTagger** (Kunin & Hugenholtz, 2010), *etc.* The **ggtree** supports visualizing the phylogenetic tree stored in the `phyloseq` object and related data can be used to

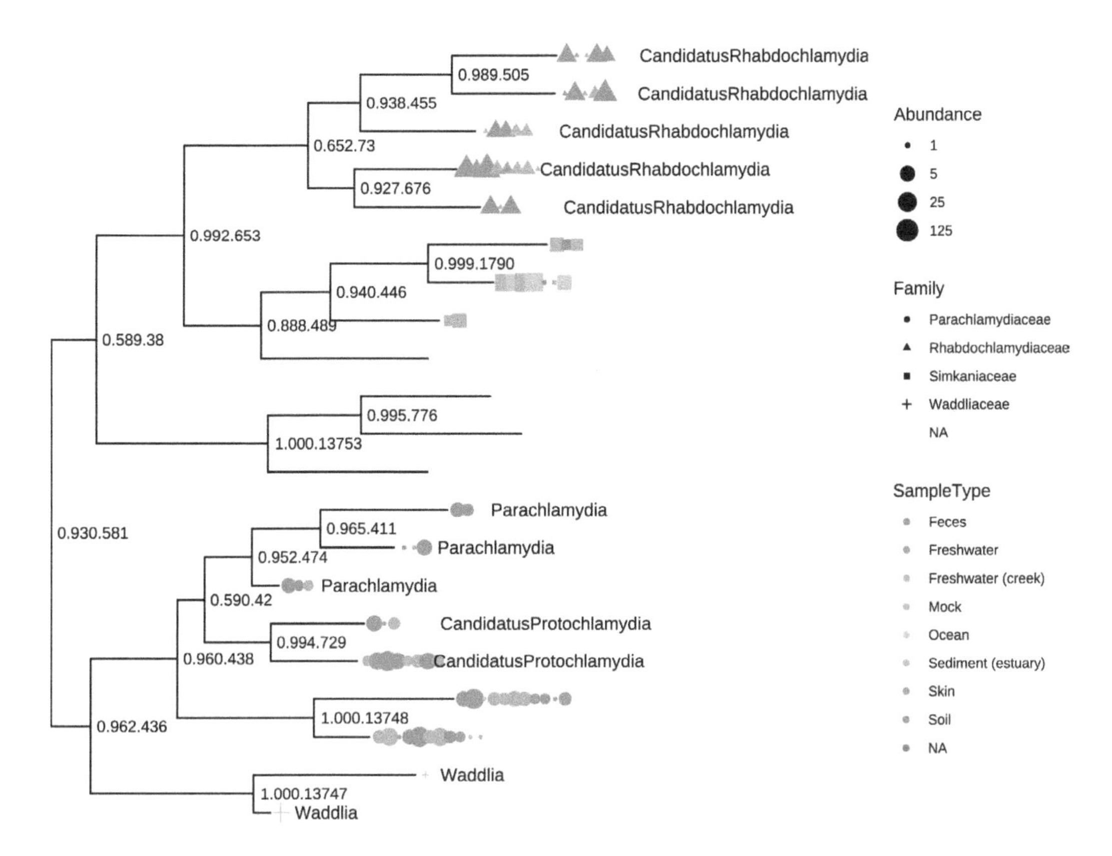

Figure 9.3: **Visualizing a `phyloseq` tree object.** This example mimics the output of the `plot_tree()` function provided in the **phyloseq** package.

annotate the tree as demonstrated in Figures 9.3 and 9.4.

```
library(phyloseq)
library(scales)

data(GlobalPatterns)
GP <- prune_taxa(taxa_sums(GlobalPatterns) > 0, GlobalPatterns)
GP.chl <- subset_taxa(GP, Phylum=="Chlamydiae")

ggtree(GP.chl) +
  geom_nodelab(aes(label=label), hjust=-.05, size=3.5) +

  geom_point(aes(x=x+hjust, color=SampleType, shape=Family,
                 size=Abundance), na.rm=TRUE) +
  geom_tiplab(aes(label=Genus), hjust=-.35) +
  scale_size_continuous(trans=log_trans(5)) +
  theme(legend.position="right") + hexpand(.4)
```

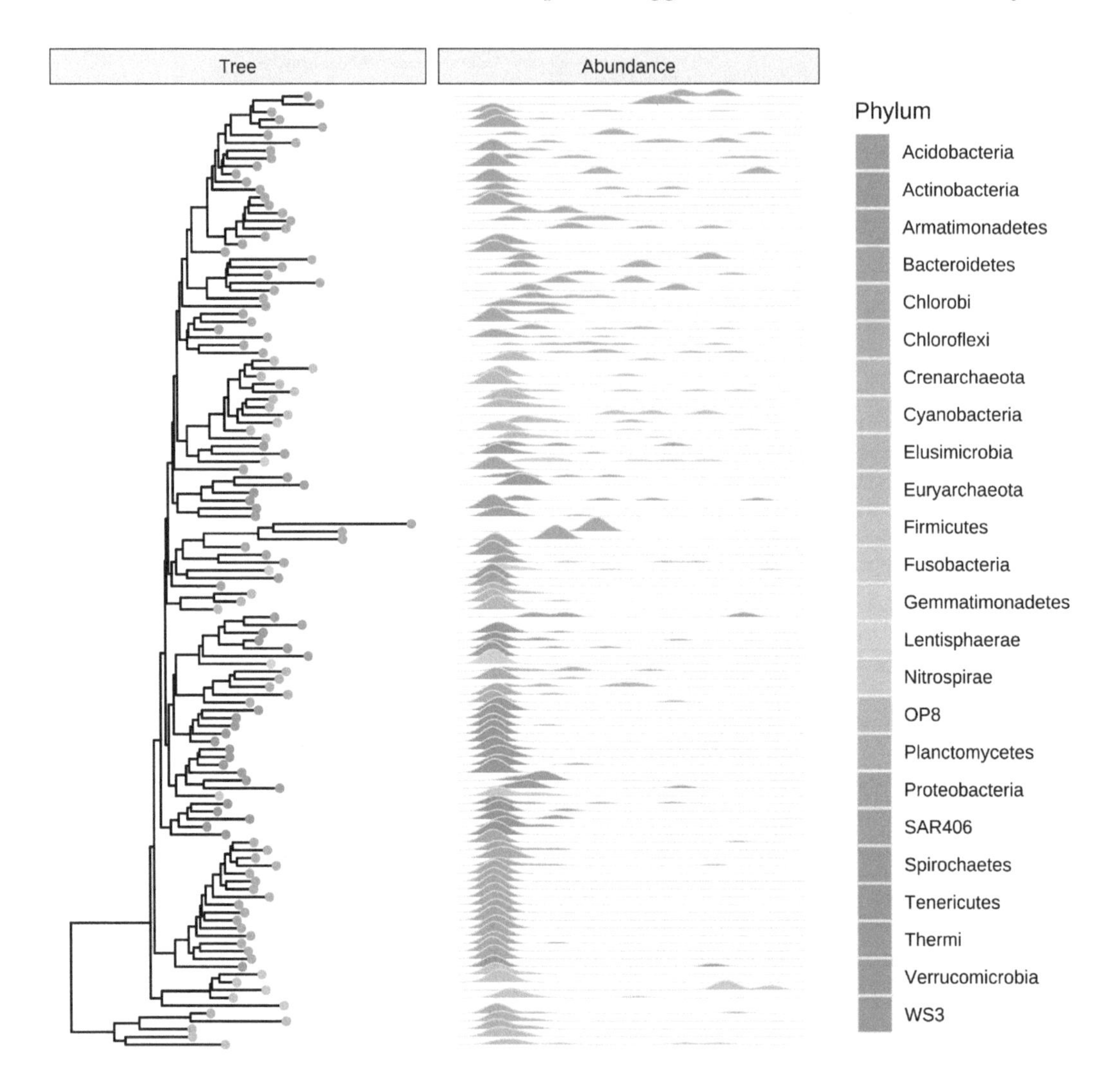

Figure 9.4: **Phylogenetic tree with OTU abundance densities.** Tips were colored by Phylum, and the corresponding abundances across different samples were visualized as density ridgelines and sorted according to the tree structure.

Figure 9.3 reproduces the output of the `phyloseq::plot_tree()` function. Users of **phyloseq** will find **ggtree** useful for visualizing microbiome data and for further annotation since **ggtree** supports high-level annotation using the grammar of graphics and can add tree data layers that are not available in **phyloseq**.

```
library(ggridges)

data("GlobalPatterns")
GP <- GlobalPatterns
GP <- prune_taxa(taxa_sums(GP) > 600, GP)
sample_data(GP)$human <- get_variable(GP, "SampleType") %in%
```

```
  c("Feces", "Skin")

mergedGP <- merge_samples(GP, "SampleType")
mergedGP <- rarefy_even_depth(mergedGP,rngseed=394582)
mergedGP <- tax_glom(mergedGP,"Order")

melt_simple <- psmelt(mergedGP) %>%
  filter(Abundance < 120) %>%
  select(OTU, val=Abundance)

ggtree(mergedGP) +
  geom_tippoint(aes(color=Phylum), size=1.5) +
  geom_facet(mapping = aes(x=val,group=label,
                           fill=Phylum),
             data = melt_simple,
             geom = geom_density_ridges,
             panel="Abundance",
             color='grey80', lwd=.3) +
  guides(color = guide_legend(ncol=1))
```

This example uses microbiome data provided in the **phyloseq** package and density ridgeline is employed to visualize species abundance data. The `geom_facet()` layer automatically re-arranges the abundance data according to the tree structure, visualizes the data using the specified `geom` function, *i.e.*, `geom_density_ridges()`, and aligns the density curves with the tree as demonstrated in Figure 9.4. Note that data stored in the `phyloseq` object is visible to `ggtree()` and can be used directly in tree visualization (`Phylum` was used to color tips and density ridgelines in this example). The source code of this example was firstly published in the supplemental file of (Yu et al., 2018).

9.2 ggtree for Dendrograms

A dendrogram is a tree diagram to display hierarchical clustering and classification/regression trees. In R, we can calculate a hierarchical clustering using the function `hclust()`.

```
hc <- hclust(dist(mtcars))
hc

##
## Call:
## hclust(d = dist(mtcars))
##
## Cluster method   : complete
## Distance         : euclidean
```

```
## Number of objects: 32
```

The `hclust` object describes the tree produced by the clustering process. It can be converted to `dendrogram` object, which stores the tree as deeply-nested lists.

```
den <- as.dendrogram(hc)
den
```

```
## 'dendrogram' with 2 branches and 32 members total, at height 425.3
```

The **ggtree** package supports most of the hierarchical clustering objects defined in the R community, including `hclust` and `dendrogram` as well as `agnes`, `diana`, and `twins` that are defined in the **cluster** package, and the `pvclust` object defined in the **pvclust** package (Table 2). Users can use `ggtree(object)` to display its tree structure, and use other layers and utilities to customize the graph and of course, add annotations to the tree.

The **ggtree** provides `layout_dendrogram()` to layout the tree top-down, and `theme_dendrogram()` to display tree height (similar to `theme_tree2()` for phylogenetic tree) as demonstrated in Figure 9.5 (see also the example in (Yu, 2020)).

```
clus <- cutree(hc, 4)
g <- split(names(clus), clus)

p <- ggtree(hc, linetype='dashed')
clades <- sapply(g, function(n) MRCA(p, n))

p <- groupClade(p, clades, group_name='subtree') + aes(color=subtree)

d <- data.frame(label = names(clus),
                cyl = mtcars[names(clus), "cyl"])

p %<+% d +
  layout_dendrogram() +
  geom_tippoint(aes(fill=factor(cyl), x=x+.5),
                size=5, shape=21, color='black') +
  geom_tiplab(aes(label=cyl), size=3, hjust=.5, color='black') +
  geom_tiplab(angle=90, hjust=1, offset=-10, show.legend=FALSE) +
  scale_color_brewer(palette='Set1', breaks=1:4) +
  theme_dendrogram(plot.margin=margin(6,6,80,6)) +
  theme(legend.position=c(.9, .6))
```

9.3 ggtree for Tree Graph

The tree graph (as an `igraph` object) can be converted to a `phylo` object using `as.phylo()` method provided in the **treeio** package (Table 2). The **ggtree** supports directly visualizing tree graph as demonstrated in Figure 9.6. Note that currently not

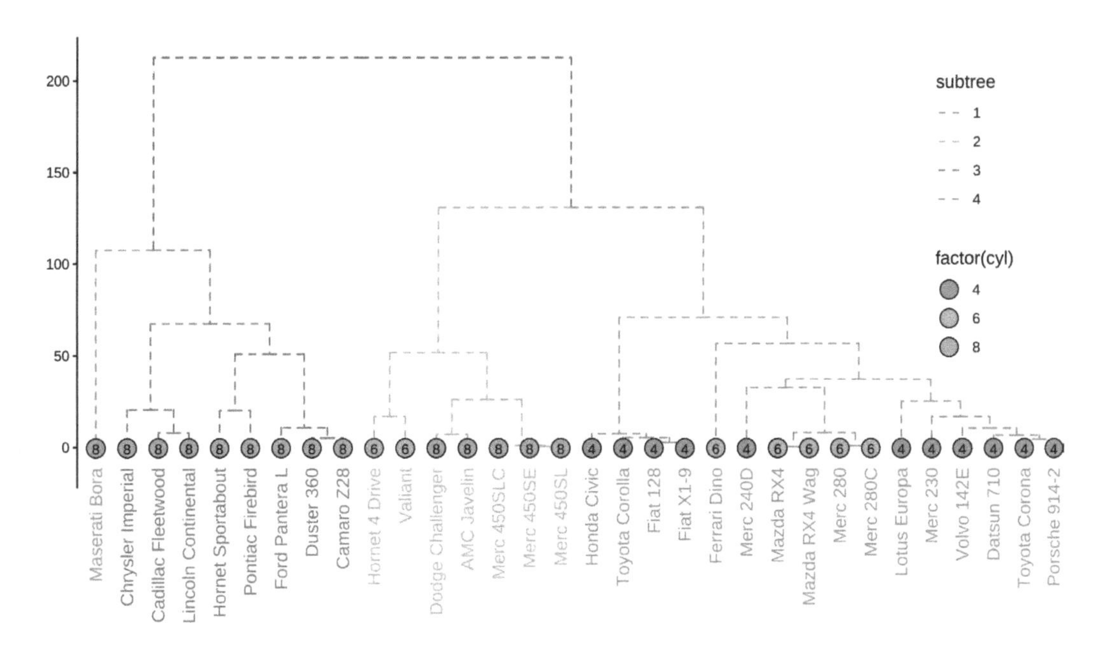

Figure 9.5: **Visualizing dendrogram.** Use `cutree()` to split the tree into several groups and `groupClade()` to assign this grouping information. The tree was displayed in the classic top-down layout with branches colored by the grouping information and the tips were colored and labeled by the number of cylinders.

all `igraph` objects can be supported by **ggtree**. Currently, it can only be supported when it is a tree graph.

```
library(igraph)
g <- graph.tree(40, 3)
arrow_size <- unit(rep(c(0, 3), times = c(27, 13)), "mm")
ggtree(g, layout='slanted', arrow = arrow(length=arrow_size)) +
  geom_point(size=5, color='steelblue', alpha=.6) +
  geom_tiplab(hjust=.5,vjust=2) + layout_dendrogram()
```

9.4 ggtree for Other Tree-like Structures

The **ggtree** package can be used to visualize any data in a hierarchical structure. Here, we use the GNI (Gross National Income) numbers in 2014 as an example. After preparing an edge list, that is a matrix or data frame that contains two columns indicating the relationship of parent and child nodes, we can use the `as.phylo()` method provided by the **treeio** package to convert the edge list to a `phylo` object. Then it can be visualized using **ggtree** with associated data. In this example, the population was used to scale the size of circle points for each country (Figure 9.7).

```
library(treeio)
library(ggplot2)
```

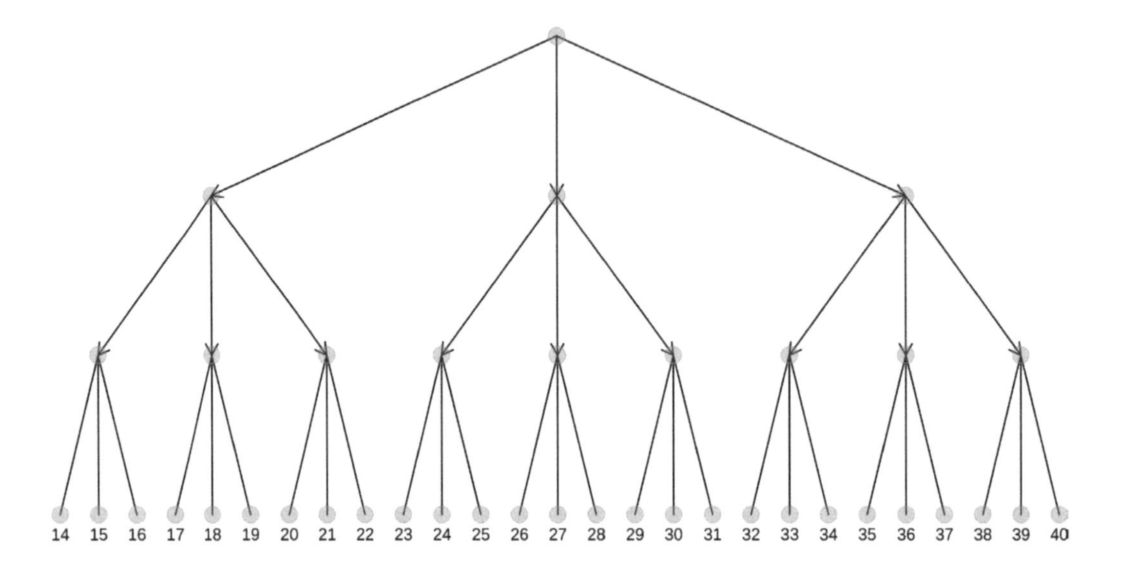

Figure 9.6: **Visualizing a tree graph.** The lines with arrows indicate the relationship between the parent node and the child node. All nodes were indicated by steelblue circle points.

```
library(ggtree)

data("GNI2014", package="treemap")
n <- GNI2014[, c(3,1)]
n[,1] <- as.character(n[,1])
n[,1] <- gsub("\\s\\(.*\\)", "", n[,1])

w <- cbind("World", as.character(unique(n[,1])))

colnames(w) <- colnames(n)
edgelist <- rbind(n, w)

y <- as.phylo(edgelist)
ggtree(y, layout='circular') %<+% GNI2014 +
    aes(color=continent) + geom_tippoint(aes(size=population), alpha=.6) +
    geom_tiplab(aes(label=country), offset=.05, size=3) +
    xlim(NA, 3)
```

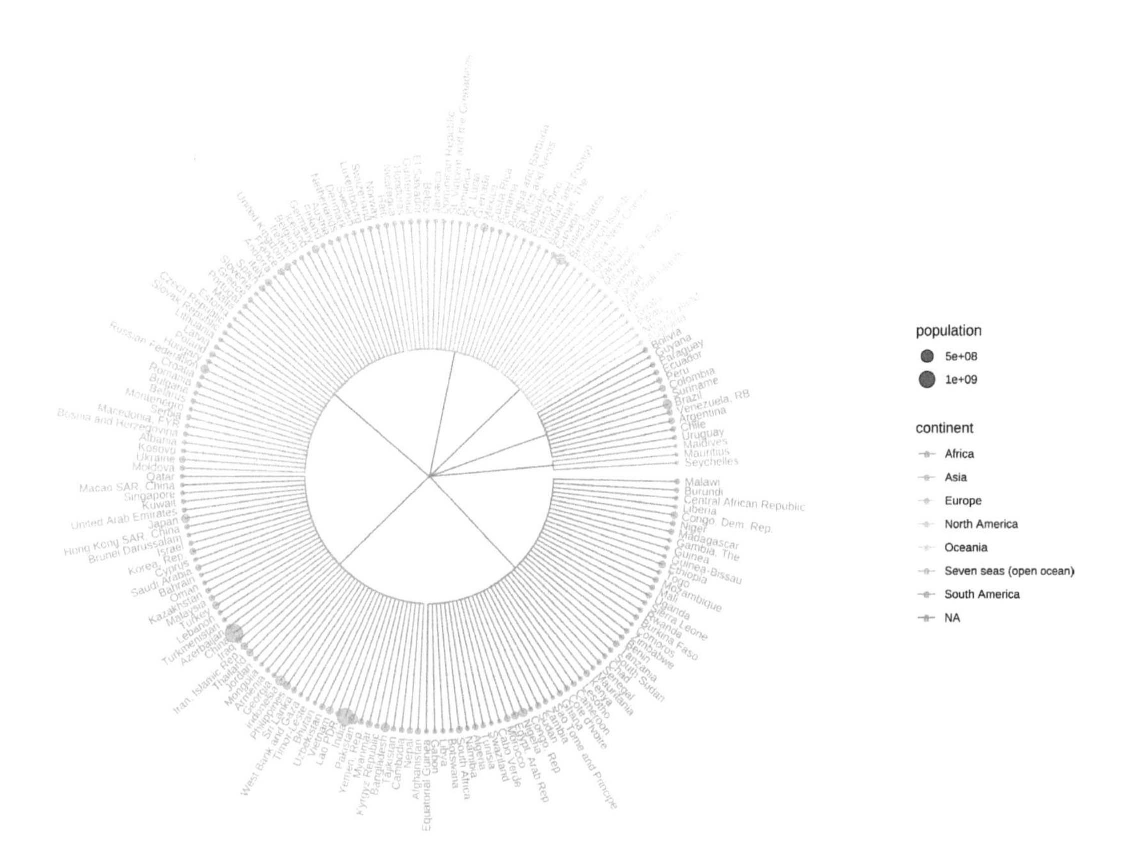

Figure 9.7: **Visualizing data in any hierarchical structure.** Hierarchical data represented as nodes connected by edges can be converted to a `phylo` object and visualized by **ggtree** to explore their relationships or other properties that are associated with the relationships.

9.5 Summary

The **ggtree** supports various tree objects defined in the R language and extension packages, which makes it very easy to integrate **ggtree** into existing pipelines. Moreover, **ggtree** allows external data integration and exploration of these data on the tree, which will greatly promote the data visualization and result in interpretation in the downstream analysis of existing pipelines. Most importantly, the support for converting edge list to a tree object enables more tree-like structures to be incorporated into the framework of **treeio** and **ggtree**. This will enable more tree-like structures and related heterogeneous data in different disciplines to be integrated and visualized through **treeio** and **ggtree**, which facilitates integrated analysis and comparative analysis to discover more systematic patterns and insights.

Part III: ggtree extensions

Chapter 10

ggtreeExtra for Presenting Data on a Circular Layout

10.1 Introduction

The **ggtree** package (Yu et al., 2017) provides programmable visualization and annotation of phylogenetic trees and other tree-like structures. It supports visualizing tree data in multiple layers or with the tree side-by-side (see also Chapter 7 and (Yu et al., 2018)). Although **ggtree** supports many layouts, the `geom_facet()` layer only works with `rectangular`, `roundrect`, `ellipse`, and `slanted` layouts to present tree data on different panels. There is no direct support in **ggtree** to present data on the outer rings of a tree in `circular`, `fan`, and `radial` layouts. To solve this issue, we developed the **ggtreeExtra** package, which allows users to align associated graph layers in outer rings of circular layout tree. In addition, it also works with a `rectangular` tree layout (Figure 10.3).

10.2 Aligning Graphs to the Tree Based on a Tree Structure

The **ggtreeExtra** package provides a layer function, `geom_fruit()`, to align graphs with the tree side-by-side. Similar to the `geom_facet()` layout described in Chapter 7, `geom_fruit()` internally re-orders the input data based on the tree structure and visualizes the data using a specified geometric layer function with user-provided aesthetic mapping and non-variable setting. The graph will be displayed on the outer ring of the tree.

The `geom_fruit()` is designed to work with most `geom` layers defined in **ggplot2** and its extensions. The position of the graph (*i.e.*, on the outer ring) is controlled by the `position` parameter, which accepts a `Position` object. The default value of the `position` parameter is `auto'` `and the`geom_fruit()`layer will guess and determine (hopefully) a` suitable position for the specified geometric layer. That means usingposition_stackx()forgeom_bar(),position_dodgex()forgeom_violin()andgeom_boxplot(), andposition_identityx()for others (*e.g.*,geom_point(),geom_tile(), *etc.*). A geometric layer that has apositionparameter should be compatible withgeom_fruit(), as it allows using position functions defined in the [**ggtreeExtra**](http://bioconductor.org/packages/ggtreeExtra) package to adjust output layer positions. Besides, thegeom_fruit()layer allows settingaxisand background grid lines for the current layer using theaxis.paramsandgrid.params‘ parameters, respectively.

The following example uses microbiome data provided in the **phyloseq** package and a boxplot is employed to visualize species abundance data. The `geom_fruit()` layer automatically rearranges the abundance data according to the circular tree structure and visualizes the data using the specific `geom` function (*i.e.*, `geom_boxplot()`). Visualizing this dataset using `geom_density_ridges()` with `geom_facet()` can be found in figure 1 of (Yu et al., 2018).

```
library(ggtreeExtra)
library(ggtree)
library(phyloseq)
library(dplyr)

data("GlobalPatterns")
GP <- GlobalPatterns
GP <- prune_taxa(taxa_sums(GP) > 600, GP)
sample_data(GP)$human <- get_variable(GP, "SampleType") %in%
                              c("Feces", "Skin")
mergedGP <- merge_samples(GP, "SampleType")
mergedGP <- rarefy_even_depth(mergedGP,rngseed=394582)
mergedGP <- tax_glom(mergedGP,"Order")

melt_simple <- psmelt(mergedGP) %>%
               filter(Abundance < 120) %>%
               select(OTU, val=Abundance)

p <- ggtree(mergedGP, layout="fan", open.angle=10) +
     geom_tippoint(mapping=aes(color=Phylum),
                   size=1.5,
                   show.legend=FALSE)
p <- rotate_tree(p, -90)

p <- p +
     geom_fruit(
         data=melt_simple,
         geom=geom_boxplot,
```

```
          mapping = aes(
                      y=OTU,
                      x=val,
                      group=label,
                      fill=Phylum,
                    ),
          size=.2,
          outlier.size=0.5,
          outlier.stroke=0.08,
          outlier.shape=21,
          axis.params=list(
                          axis       = "x",
                          text.size  = 1.8,
                          hjust      = 1,
                          vjust      = 0.5,
                          nbreak     = 3,
                      ),
          grid.params=list()
      )

p <- p +
      scale_fill_discrete(
          name="Phyla",
          guide=guide_legend(keywidth=0.8, keyheight=0.8, ncol=1)
      ) +
      theme(
          legend.title=element_text(size=9),
          legend.text=element_text(size=7)
      )
p
```

10.3 Aligning Multiple Graphs to the Tree for Multi-dimensional Data

We are able to add multiple `geom_fruit()` layers to a tree and the circular layout is indeed more compact and efficient for multi-dimensional data. This example reproduces figure 2 of (Morgan et al., 2013). The data is provided by GraPhlAn (Asnicar et al., 2015), which contained the relative abundance of the microbiome at different body sites. This example demonstrates the ability to add multiple layers (heat map and bar plot) to present different types of data (Figure 10.2).

```
library(ggtreeExtra)
library(ggtree)
library(treeio)
```

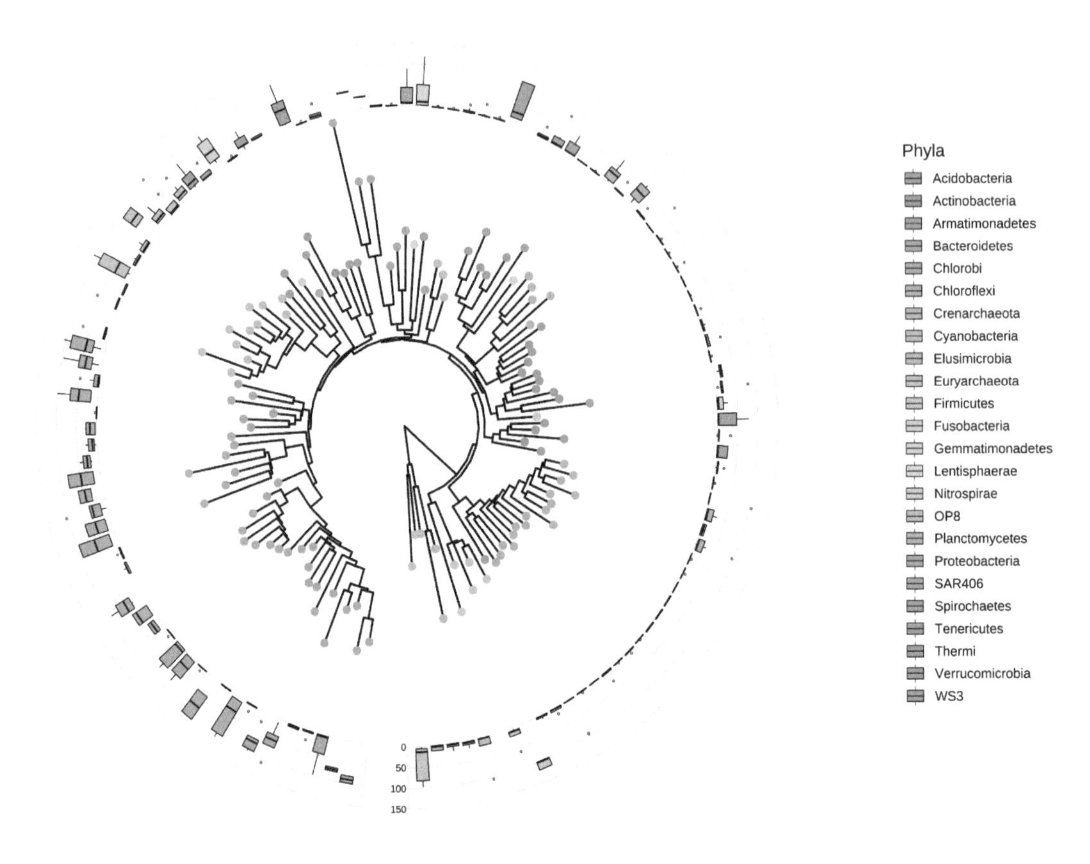

Figure 10.1: **Phylogenetic tree with OTU abundance distribution**. Species abundance distribution was aligned to the tree and visualized as boxplots. The Phylum information was used to color symbolic points on the tree and also species abundance distributions.

```
library(tidytree)
library(ggstar)
library(ggplot2)
library(ggnewscale)
library(TDbook)

# load data from TDbook, including tree_hmptree,
# df_tippoint (the abundance and types of microbes),
# df_ring_heatmap (the abundance of microbes at different body sites),
# and df_barplot_attr (the abundance of microbes of greatest prevalence)
tree <- tree_hmptree
dat1 <- df_tippoint
dat2 <- df_ring_heatmap
dat3 <- df_barplot_attr

# adjust the order
dat2$Sites <- factor(dat2$Sites,
                    levels=c("Stool (prevalence)", "Cheek (prevalence)",
```

```
                                "Plaque (prevalence)","Tongue (prevalence)",
                                "Nose (prevalence)", "Vagina (prevalence)",
                                "Skin (prevalence)"))
dat3$Sites <- factor(dat3$Sites,
                     levels=c("Stool (prevalence)", "Cheek (prevalence)",
                              "Plaque (prevalence)", "Tongue (prevalence)",
                              "Nose (prevalence)", "Vagina (prevalence)",
                              "Skin (prevalence)"))
# extract the clade label information. Because some nodes of tree are
# annotated to genera, which can be displayed with high light using ggtree.
nodeids <- nodeid(tree, tree$node.label[nchar(tree$node.label)>4])
nodedf <- data.frame(node=nodeids)
nodelab <- gsub("[\\.0-9]", "", tree$node.label[nchar(tree$node.label)>4])
# The layers of clade and hightlight
poslist <- c(1.6, 1.4, 1.6, 0.8, 0.1, 0.25, 1.6, 1.6, 1.2, 0.4,
             1.2, 1.8, 0.3, 0.8, 0.4, 0.3, 0.4, 0.4, 0.4, 0.6,
             0.3, 0.4, 0.3)
labdf <- data.frame(node=nodeids, label=nodelab, pos=poslist)

# The circular layout tree.
p <- ggtree(tree, layout="fan", size=0.15, open.angle=5) +
     geom_hilight(data=nodedf, mapping=aes(node=node),
                  extendto=6.8, alpha=0.3, fill="grey", color="grey50",
                  size=0.05) +
     geom_cladelab(data=labdf,
                   mapping=aes(node=node,
                               label=label,
                               offset.text=pos),
                   hjust=0.5,
                   angle="auto",
                   barsize=NA,
                   horizontal=FALSE,
                   fontsize=1.4,
                   fontface="italic"
                   )

p <- p %<+% dat1 + geom_star(
                        mapping=aes(fill=Phylum, starshape=Type, size=Size),
                        position="identity",starstroke=0.1) +
        scale_fill_manual(values=c("#FFC125","#87CEFA","#7B68EE","#808080",
                                   "#800080", "#9ACD32","#D15FEE","#FFC0CB",
                                   "#EE6A50","#8DEEEE", "#006400","#800000",
                                   "#B0171F","#191970"),
                          guide=guide_legend(keywidth = 0.5,
                                             keyheight = 0.5, order=1,
                                             override.aes=list(starshape=15)),
                          na.translate=FALSE)+
        scale_starshape_manual(values=c(15, 1),
                          guide=guide_legend(keywidth = 0.5,
                                             keyheight = 0.5, order=2),
```

```
                        na.translate=FALSE)+
    scale_size_continuous(range = c(1, 2.5),
                        guide = guide_legend(keywidth = 0.5,
                                    keyheight = 0.5, order=3,
                                    override.aes=list(starshape=15)))

p <- p + new_scale_fill() +
        geom_fruit(data=dat2, geom=geom_tile,
                mapping=aes(y=ID, x=Sites, alpha=Abundance, fill=Sites),
                color = "grey50", offset = 0.04,size = 0.02)+
        scale_alpha_continuous(range=c(0, 1),
                            guide=guide_legend(keywidth = 0.3,
                                            keyheight = 0.3, order=5)) +
        geom_fruit(data=dat3, geom=geom_bar,
                    mapping=aes(y=ID, x=HigherAbundance, fill=Sites),
                    pwidth=0.38,
                    orientation="y",
                    stat="identity",
        ) +
        scale_fill_manual(values=c("#0000FF","#FFA500","#FF0000",
                            "#800000", "#006400","#800080","#696969"),
                        guide=guide_legend(keywidth = 0.3,
                                    keyheight = 0.3, order=4))+
        geom_treescale(fontsize=2, linesize=0.3, x=4.9, y=0.1) +
        theme(legend.position=c(0.93, 0.5),
              legend.background=element_rect(fill=NA),
              legend.title=element_text(size=6.5),
              legend.text=element_text(size=4.5),
              legend.spacing.y = unit(0.02, "cm"),
            )
p
```

The shape of the tip points indicates the types of microbes (commensal microbes or potential pathogens). The transparency of the heatmap indicates the abundance of the microbes, and the colors of the heatmap indicate different sites of the human body. The bar plot indicates the relative abundance of the most prevalent species at the body sites. The node labels contain taxonomy information in this example, and the information was used to highlight and label corresponding clades using `geom_hilight()` and `geom_cladelab()`, respectively.

The `geom_fruit()` layer supports rectangular layout. Users can either add a `geom_fruit()` layer to a rectangular tree (e.g., `ggtree(tree_object) + geom_fruit(...)`) or use `layout_rectangular()` to transform a circular layout tree to a rectangular layout tree as demonstrated in Figure 10.3.

```
p + layout_rectangular() +
    theme(legend.position=c(.05, .7))
```

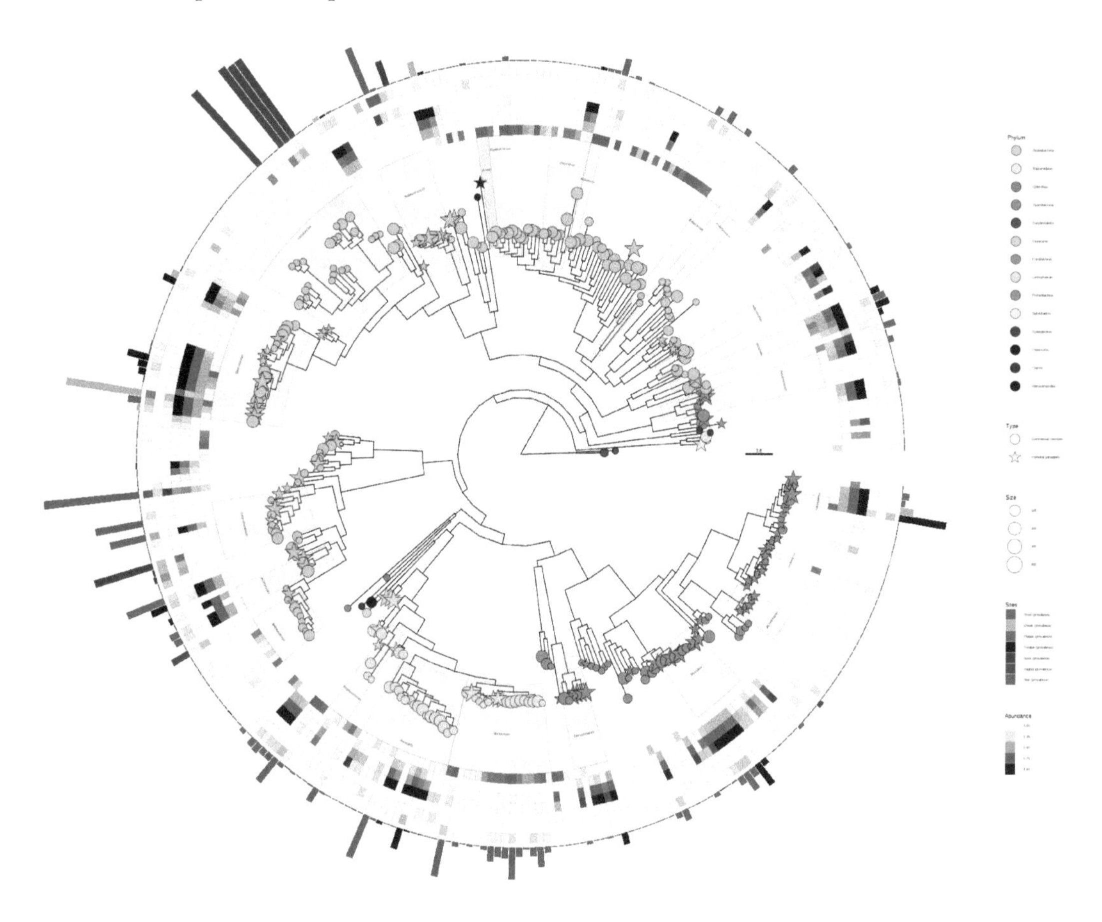

Figure 10.2: **Presenting microbiome data (abundance and location) on a phylogenetic tree.** The tree was annotated with symbolic points, highlighted clades, and clade labels. Two `geom_fruit()` layers were used to visualize location and abundance information.

10.4 Examples for Population Genetics

The **ggtree** (Yu et al., 2017) and **ggtreeExtra** packages are designed as general tools and can be applied to many research fields, such as infectious disease epidemiology, metagenome, population genetics, evolutionary biology, and ecology. We have introduced examples for metagenome research (Figure 10.1 and Figure 10.2). In this session, we present examples for population genetics by reproducing figure 4 of (Chow et al., 2020) and figure 1 of (Wong et al., 2015).

```
library(ggtree)
library(ggtreeExtra)
library(ggplot2)
library(ggnewscale)
library(reshape2)
library(dplyr)
library(tidytree)
```

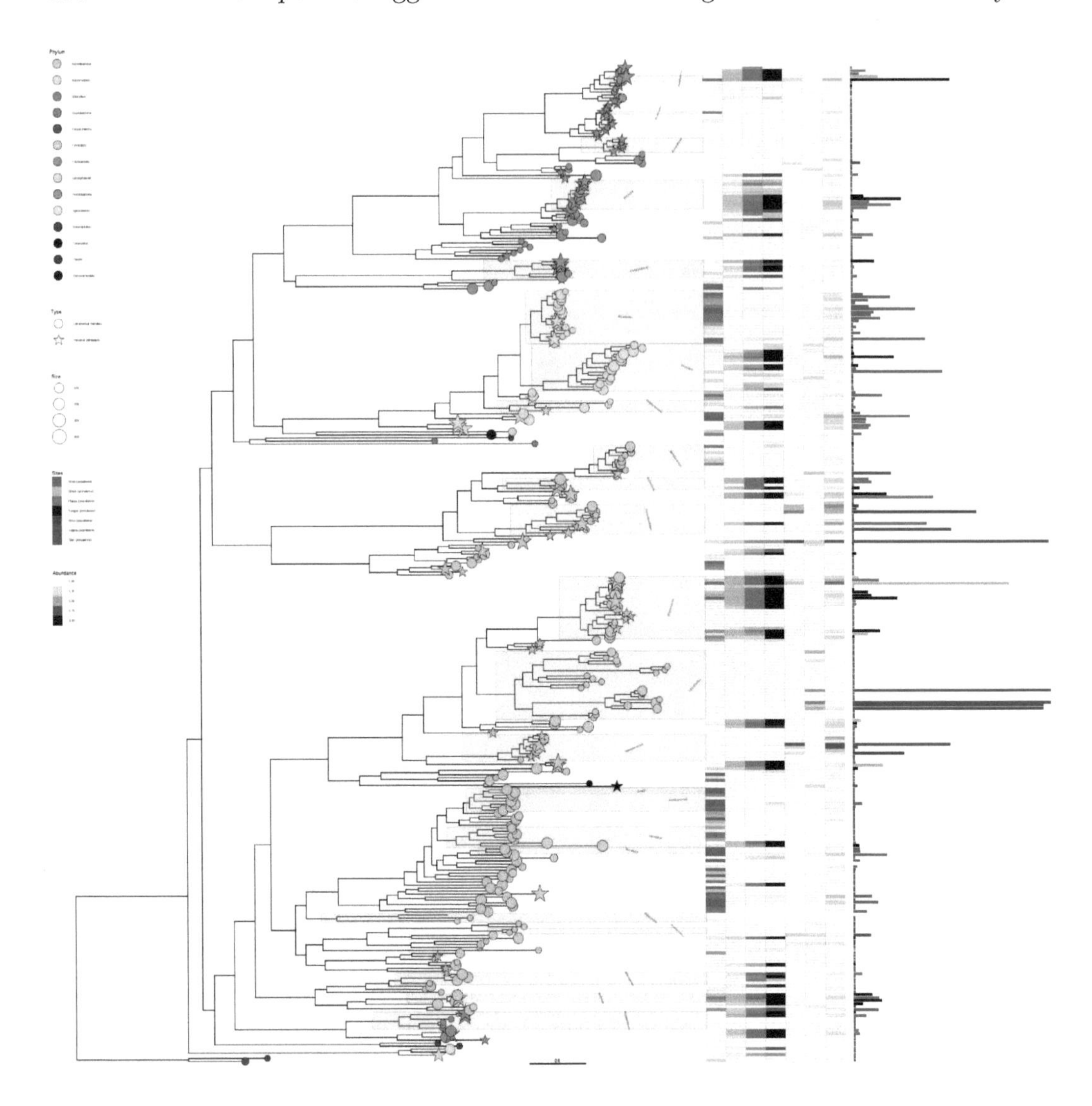

Figure 10.3: **Illustration of using `geom_fruit()` in rectangular tree layout.** The figure was produced by transforming Figure 10.2 using the rectangular layout. Transforming a rectangular layout tree to a circular layout tree is also supported.

```
library(ggstar)
library(TDbook)

# load tr and dat from the TDbook package
dat <- df_Candidaauris_data
tr <- tree_Candidaauris

countries <- c("Canada", "United States",
               "Colombia", "Panama",
               "Venezuela", "France",
```

```
                "Germany", "Spain",
                "UK", "India",
                "Israel", "Pakistan",
                "Saudi Arabia", "United Arab Emirates",
                "Kenya", "South Africa",
                "Japan", "South Korea",
                "Australia")
# For the tip points
dat1 <- dat %>% select(c("ID", "COUNTRY", "COUNTRY__colour"))
dat1$COUNTRY <- factor(dat1$COUNTRY, levels=countries)
COUNTRYcolors <- dat1[match(countries,dat$COUNTRY),"COUNTRY__colour"]

# For the heatmap layer
dat2 <- dat %>% select(c("ID", "FCZ", "AMB", "MCF"))
dat2 <- melt(dat2,id="ID", variable.name="Antifungal", value.name="type")
dat2$type <- paste(dat2$Antifungal, dat2$type)
dat2$type[grepl("Not_", dat2$type)] = "Susceptible"
dat2$Antifungal <- factor(dat2$Antifungal, levels=c("FCZ", "AMB", "MCF"))
dat2$type <- factor(dat2$type,
                    levels=c("FCZ Resistant",
                            "AMB Resistant",
                            "MCF Resistant",
                            "Susceptible"))

# For the points layer
dat3 <- dat %>% select(c("ID", "ERG11", "FKS1")) %>%
        melt(id="ID", variable.name="point", value.name="mutation")
dat3$mutation <- paste(dat3$point, dat3$mutation)
dat3$mutation[grepl("WT", dat3$mutation)] <- NA
dat3$mutation <- factor(dat3$mutation,
                        levels=c("ERG11 Y132F", "ERG11 K143R",
                                 "ERG11 F126L", "FKS1 S639Y/P/F"))

# For the clade group
dat4 <- dat %>% select(c("ID", "CLADE"))
dat4 <- aggregate(.~CLADE, dat4, FUN=paste, collapse=",")
clades <- lapply(dat4$ID, function(x){unlist(strsplit(x,split=","))})
names(clades) <- dat4$CLADE

tr <- groupOTU(tr, clades, "Clade")
Clade <- NULL
p <- ggtree(tr=tr, layout="fan", open.angle=15,size=0.2, aes(colour=Clade))+
     scale_colour_manual(
         values=c("black","#69B920","#9C2E88","#F74B00","#60C3DB"),
         labels=c("","I", "II", "III", "IV"),
         guide=guide_legend(keywidth=0.5,
                            keyheight=0.5,
                            order=1,
                            override.aes=list(linetype=c("0"=NA,
                                                         "Clade1"=1,
```

```
                                                         "Clade2"=1,
                                                         "Clade3"=1,
                                                         "Clade4"=1
                                                        )
                                              )
                           )
     ) +
     new_scale_colour()

p1 <- p %<+% dat1 +
     geom_tippoint(aes(colour=COUNTRY),
                 alpha=0) +
     geom_tiplab(aes(colour=COUNTRY),
                 align=TRUE,
                 linetype=3,
                 size=1,
                 linesize=0.2,
                 show.legend=FALSE
                 ) +
     scale_colour_manual(
         name="Country labels",
         values=COUNTRYcolors,
         guide=guide_legend(keywidth=0.5,
                            keyheight=0.5,
                            order=2,
                            override.aes=list(size=2,alpha=1))
     )

p2 <- p1 +
      geom_fruit(
          data=dat2,
          geom=geom_tile,
          mapping=aes(x=Antifungal, y=ID, fill=type),
          width=0.1,
          color="white",
          pwidth=0.1,
          offset=0.15
      ) +
      scale_fill_manual(
           name="Antifungal susceptibility",
           values=c("#595959", "#B30000", "#020099", "#E6E6E6"),
           na.translate=FALSE,
           guide=guide_legend(keywidth=0.5,
                              keyheight=0.5,
                              order=3
                             )
      ) +
      new_scale_fill()

p3 <- p2 +
```

```
    geom_fruit(
        data=dat3,
        geom=geom_star,
        mapping=aes(x=mutation, y=ID, fill=mutation, starshape=point),
        size=1,
        starstroke=0,
        pwidth=0.1,
        inherit.aes = FALSE,
        grid.params=list(
                        linetype=3,
                        size=0.2
                    )

    ) +
    scale_fill_manual(
        name="Point mutations",
        values=c("#329901", "#0600FF", "#FF0100", "#9900CC"),
        guide=guide_legend(keywidth=0.5, keyheight=0.5, order=4,
                           override.aes=list(
                                   starshape=c("ERG11 Y132F"=15,
                                               "ERG11 K143R"=15,
                                               "ERG11 F126L"=15,
                                               "FKS1 S639Y/P/F"=1),
                                   size=2)
                           ),
        na.translate=FALSE,
    ) +
    scale_starshape_manual(
        values=c(15, 1),
        guide="none"
    ) +
    theme(
        legend.background=element_rect(fill=NA),
        legend.title=element_text(size=7),
        legend.text=element_text(size=5.5),
        legend.spacing.y = unit(0.02, "cm")
    )
p3
```

In this example, Figure 10.4 shows the phylogenetic tree annotated with different colors to display different clades. The external heatmaps present the susceptibility to fluconazole (FCZ), amphotericin B (AMB), and micafungin (MCF). The external points display the point mutations in lanosterol 14-alpha-demethylase ERG11 (Y132F, K143R, and F126L) and beta-1,3-D-glucan synthase FKS1 (S639Y/P/F) associated with resistance (Chow et al., 2020).

```
library(ggtreeExtra)
library(ggtree)
library(ggplot2)
```

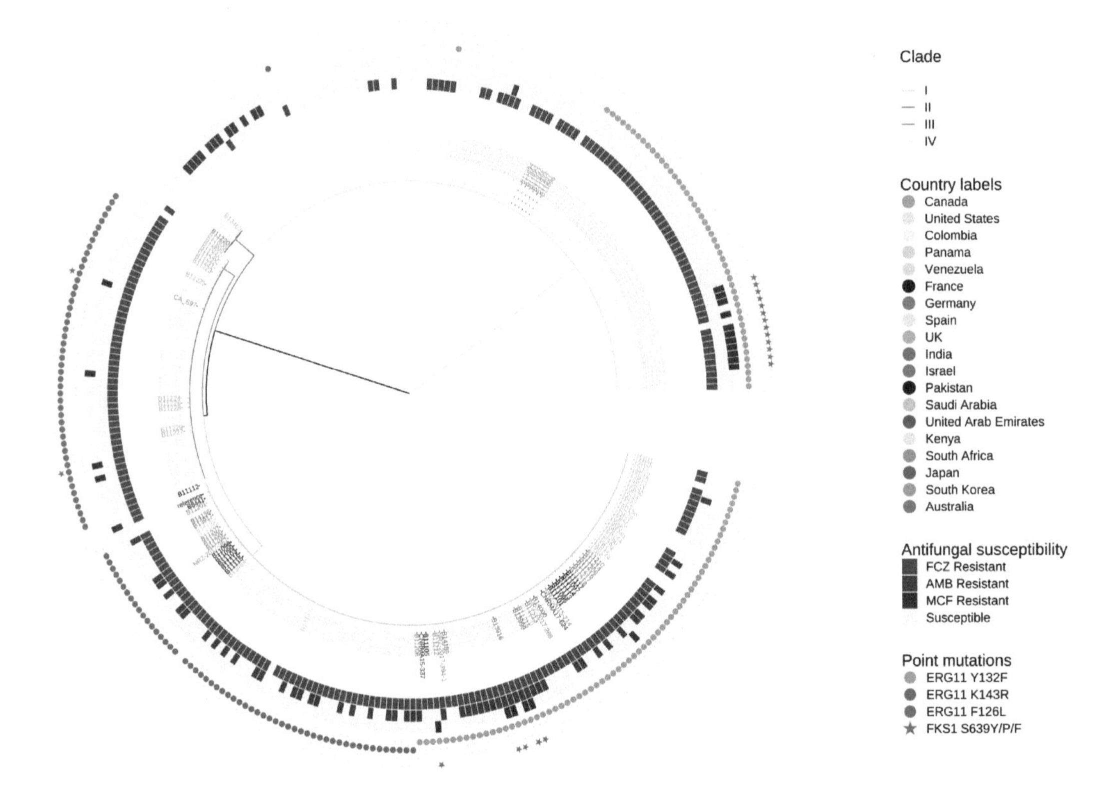

Figure 10.4: **Antifungal susceptibility and point mutations in drug targets in *Candida Auris*.**

```
library(ggnewscale)
library(treeio)
library(tidytree)
library(dplyr)
library(ggstar)
library(TDbook)

# load tree_NJIDqgsS and df_NJIDqgsS from TDbook
tr <- tree_NJIDqgsS
metada <- df_NJIDqgsS
metadata <- metada %>%
            select(c("id", "country", "country__colour",
                    "year", "year__colour", "haplotype"))
metadata$haplotype[nchar(metadata$haplotype) == 0] <- NA

countrycolors <- metada %>%
                 select(c("country", "country__colour")) %>%
                 distinct()
```

```
yearcolors <- metada %>%
              select(c("year", "year__colour")) %>%
              distinct()
yearcolors <- yearcolors[order(yearcolors$year, decreasing=TRUE),]

metadata$country <- factor(metadata$country,
                          levels=countrycolors$country)
metadata$year <- factor(metadata$year, levels=yearcolors$year)

p <- ggtree(tr, layout="fan", open.angle=15, size=0.1)

p <- p %<+% metadata

p1 <-p +
     geom_tippoint(
         mapping=aes(colour=country),
         size=1.5,
         stroke=0,
         alpha=0.4
     ) +
     scale_colour_manual(
         name="Country",
         values=countrycolors$country__colour,
         guide=guide_legend(keywidth=0.3,
                            keyheight=0.3,
                            ncol=2,
                            override.aes=list(size=2,alpha=1),
                            order=1)
     ) +
     theme(
         legend.title=element_text(size=5),
         legend.text=element_text(size=4),
         legend.spacing.y = unit(0.02, "cm")
     )

p2 <-p1 +
     geom_fruit(
         geom=geom_star,
         mapping=aes(fill=haplotype),
         starshape=26,
         color=NA,
         size=2,
         starstroke=0,
         offset=0,
```

```
    ) +
    scale_fill_manual(
        name="Haplotype",
        values=c("red"),
        guide=guide_legend(
                    keywidth=0.3,
                    keyheight=0.3,
                    order=3
              ),
        na.translate=FALSE
    )

p3 <-p2 +
    new_scale_fill() +
    geom_fruit(
        geom=geom_tile,
        mapping=aes(fill=year),
        width=0.002,
        offset=0.1
    ) +
    scale_fill_manual(
        name="Year",
        values=yearcolors$year__colour,
        guide=guide_legend(keywidth=0.3, keyheight=0.3, ncol=2,
                           order=2)
    ) +
    theme(
          legend.title=element_text(size=6),
          legend.text=element_text(size=4.5),
          legend.spacing.y = unit(0.02, "cm")
          )
p3
```

Figure 10.5 is a rooted maximum-likelihood tree of *S. Typhi* inferred from 22,145 SNPs (Wong et al., 2015), the colors of the tip points represent the geographical origin of the isolates, and the red symbolic points indicate the haplotype of H58 lineage. The color of the external heatmap indicates the years of isolation (Wong et al., 2015).

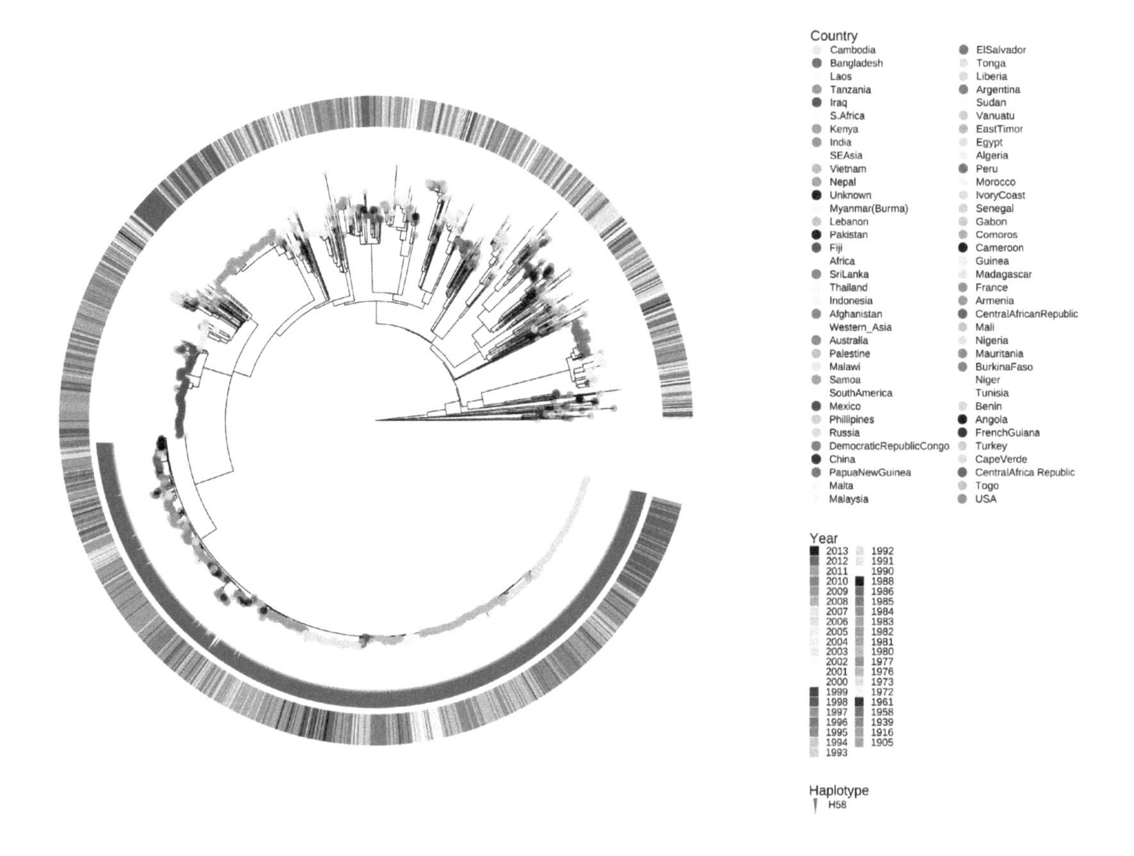

Figure 10.5: **Population structure of the 1,832 *S. Typhi* isolates.**

10.5 Summary

Compared to `geom_facet()`, `geom_fruit()` layer provided in **ggtreeExtra** is a better implementation of Method 2 proposed by (Yu et al., 2018). The `geom_facet()` and `geom_fruit()` have the same design philosophy and have a similar user interface. They rely on other geometric layers to visualize the tree-associated data. These dependent layers are provided by **ggplot2** and its extension packages, including **ggtree**. As more and more layers are implemented by the **ggplot2** community, the types of data and graphics that `geom_facet()` and `geom_fruit()` can present will also increase.

Chapter 11

Other ggtree Extensions

The **ggtree** package is a general package for visualizing tree structures and associated data. If you have some special requirements that are not directly provided by **ggtree**, you may need to use one of the extension packages built on top of **ggtree**. For example, the **RevGadgets** package for visualizing the output of the **RevBayes**, the **sitePath** package for visualizing fixation events on phylogenetic pathways, and the **enrichplot** package for visualizing hierarchical structure of the enriched pathways.

```
rp <- BiocManager::repositories()
db <- utils::available.packages(repo=rp)
x <- tools::package_dependencies('ggtree', db=db,
                                 which = c("Depends", "Imports"),
                                 reverse=TRUE)
print(x)
```

```
## $ggtree
##  [1] "enrichplot"        "ggtreeExtra"
##  [3] "LymphoSeq"         "miaViz"
##  [5] "microbiomeMarker"  "MicrobiotaProcess"
##  [7] "philr"             "singleCellTK"
##  [9] "sitePath"          "systemPipeTools"
## [11] "tanggle"           "treekoR"
```

There are 12 packages in CRAN or Bioconductor that depend on or import **ggtree** and several packages on GitHub that extend **ggtree**. Here we briefly introduce some extension packages, including **MicrobiotaProcess** and **tanggle**.

11.1 Taxonomy Annotation Using MicrobiotaProcess

The **MicrobiotaProcess** package provides a LEfSe-like algorithm (Segata et al., 2011) to discover microbiome biomarkers by comparing taxon abundance between different classes. It provides several methods to visualize the analysis result. The ggdiffclade is developed based on **ggtree** (Yu et al., 2017). In addition to the diff_analysis() result, it also supports a data frame that contains a hierarchical relationship (*e.g.*, taxonomy annotation or KEGG annotation) with another data frame that contains taxa and factor information and/or pvalue. The following example demonstrates how to use data frames (*i.e.*, analysis results) to visualize the differential taxonomy tree. More details can be found on the vignette of the **MicrobiotaProcess** package.

```
library(MicrobiotaProcess)
library(ggplot2)
library(TDbook)

# load `df_difftax` and `df_difftax_info` from TDbook
taxa <- df_alltax_info
dt <- df_difftax

ggdiffclade(obj=taxa,
            nodedf=dt,
            factorName="DIAGNOSIS",
            layout="radial",
            skpointsize=0.6,
            cladetext=2,
            linewd=0.2,
            taxlevel=3,
          # This argument is to remove the branch of unknown taxonomy.
            reduce=TRUE) +
     scale_fill_manual(values=c("#00AED7", "#009E73"))+
     guides(color = guide_legend(keywidth = 0.1, keyheight = 0.6,
                                 order = 3,ncol=1)) +
     theme(panel.background=element_rect(fill=NA),
           legend.position="right",
           plot.margin=margin(0,0,0,0),
           legend.spacing.y=unit(0.02, "cm"),
           legend.title=element_text(size=7.5),
           legend.text=element_text(size=5.5),
           legend.box.spacing=unit(0.02,"cm")
        )
```

The data frame of this example is from the analysis result of diff_analysis() using public datasets (Kostic et al., 2012). The colors represent the features enriched in the relevant class groups. The size of circle points represents the -log10(pvalue), *i.e.*, a larger point indicates a greater significance. In Figure 11.1, we can find that *Fusobacterium* sequences were enriched in carcinomas, while Firmicutes, Bacteroides, and Clostridiales were greatly reduced in tumors. These results were consistent with the original article (Kostic et al., 2012). The species of *Campylobacter* has been proven to be associated with colorectal cancer (Amer et al., 2017; He et al., 2019;

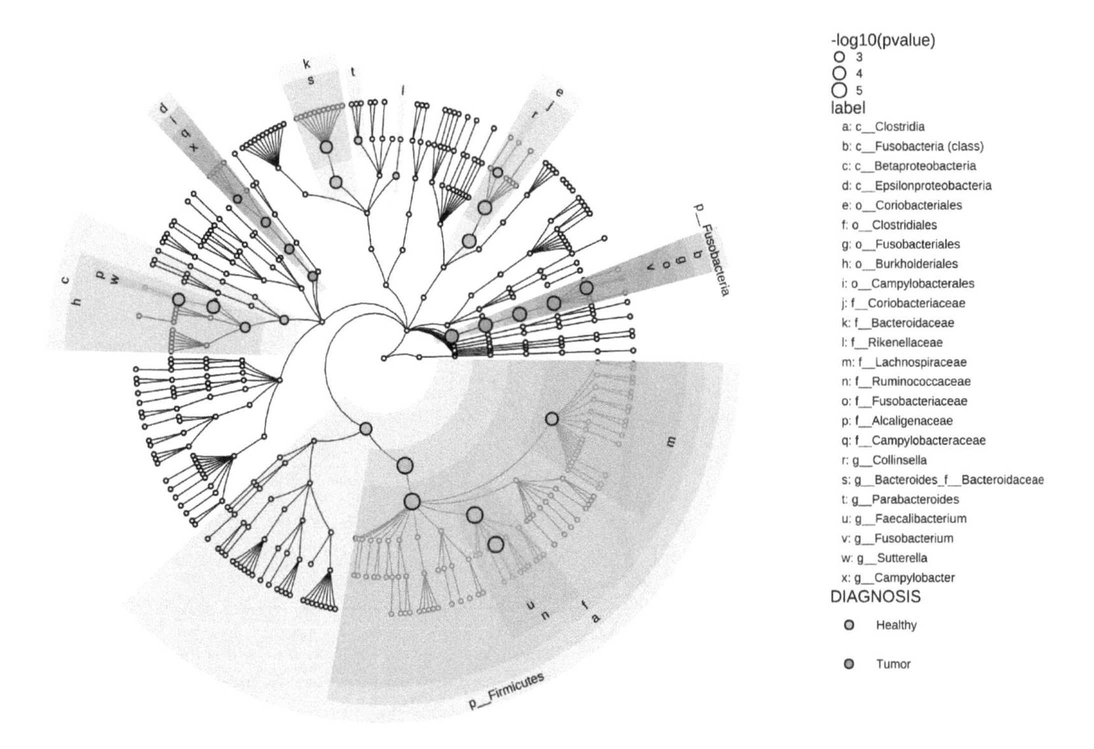

Figure 11.1: **Visualize differential taxonomy clade.**

Wu et al., 2013). We can find in Figure 11.1 that *Campylobacter* was enriched in tumors, while its relative abundance is lower than *Fusobacterium*.

11.2 Visualizing Phylogenetic Network Using Tanggle

The **tanggle** package provides functions to display a split network. It extends the **ggtree** package (Yu et al., 2017) to allow the visualization of phylogenetic networks (Figure 11.2).

```
library(ggplot2)
library(ggtree)
library(tanggle)

file <- system.file("extdata/trees/woodmouse.nxs", package = "phangorn")
Nnet <- phangorn::read.nexus.networx(file)

ggsplitnet(Nnet) +
    geom_tiplab2(aes(color=label), hjust=-.1)+
    geom_nodepoint(color='firebrick', alpha=.4) +
    scale_color_manual(values=rainbow(15)) +
    theme(legend.position="none") +
    ggexpand(.1) + ggexpand(.1, direction=-1)
```

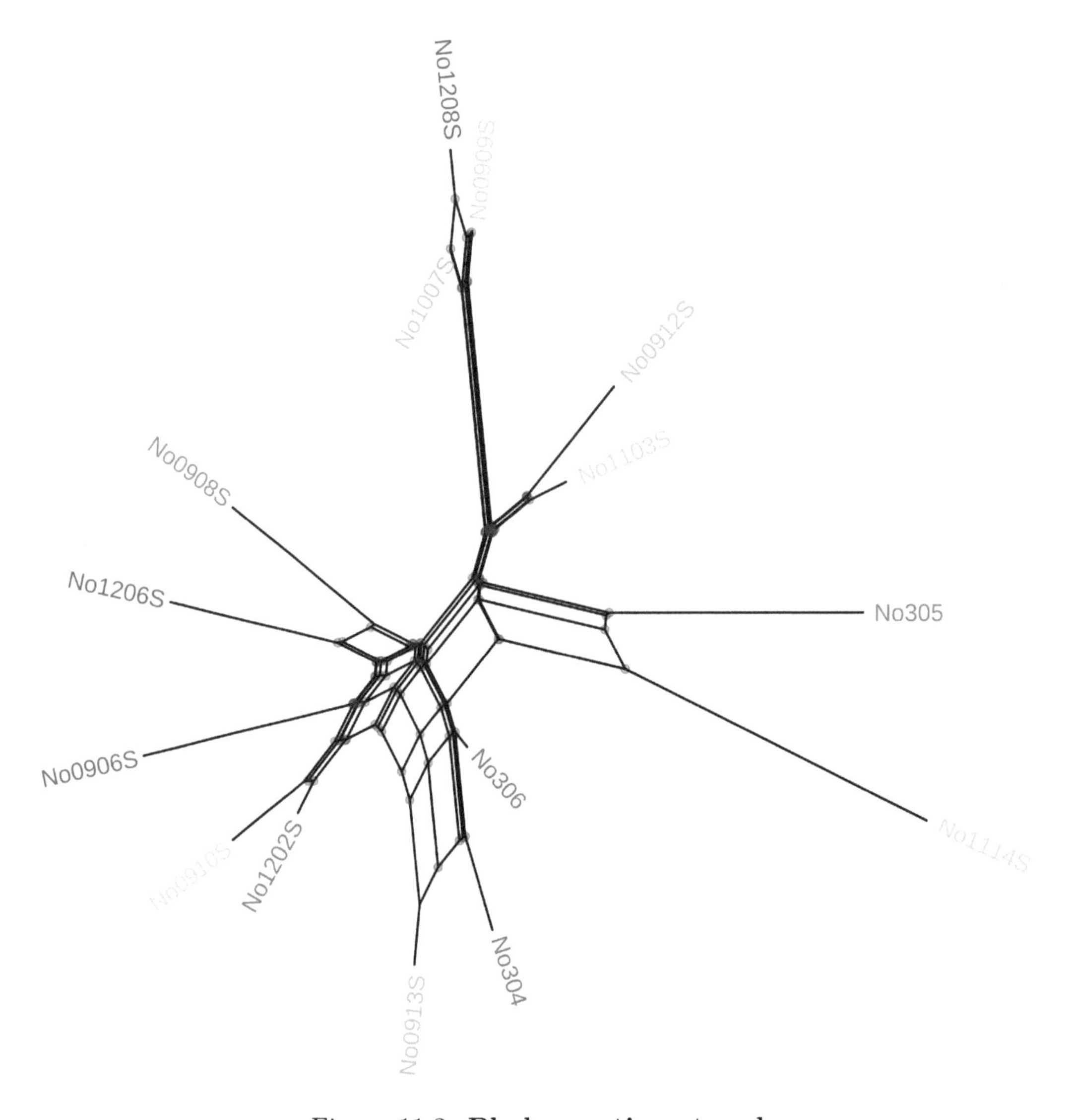

Figure 11.2: **Phylogenetic network**.

11.3 Summary

The **ggtree** is designed to support the grammar of graphics, allowing users to quickly explore phylogenetic data through visualization. When users have special needs and **ggtree** does not provide them, it is highly recommended to develop extension packages to implement these missing functions. This is a good mechanism, and we also hope that **ggtree** users can become a **ggtree** community. In this way, more functions for special needs can be developed and shared among users. Everyone will benefit from it, and it's exciting that this is happening.

Part IV: Miscellaneous topics

Chapter 12

ggtree Utilities

12.1 Facet Utilities

12.1.1 facet_widths

Adjusting relative widths of facet panels is a common requirement, especially for using `geom_facet()` to visualize a tree with associated data. However, this is not supported by the **ggplot2** package. To address this issue, **ggtree** provides the `facet_widths()` function and it works with both `ggtree` and `ggplot` objects.

```
library(ggplot2)
library(ggtree)
library(reshape2)

set.seed(123)
tree <- rtree(30)

p <- ggtree(tree, branch.length = "none") +
    geom_tiplab() + theme(legend.position='none')

a <- runif(30, 0,1)
b <- 1 - a
df <- data.frame(tree$tip.label, a, b)
df <- melt(df, id = "tree.tip.label")

p2 <- p + geom_facet(panel = 'bar', data = df, geom = geom_bar,
                mapping = aes(x = value, fill = as.factor(variable)),
                orientation = 'y', width = 0.8, stat='identity') +
        xlim_tree(9)

facet_widths(p2, widths = c(1, 2))
```

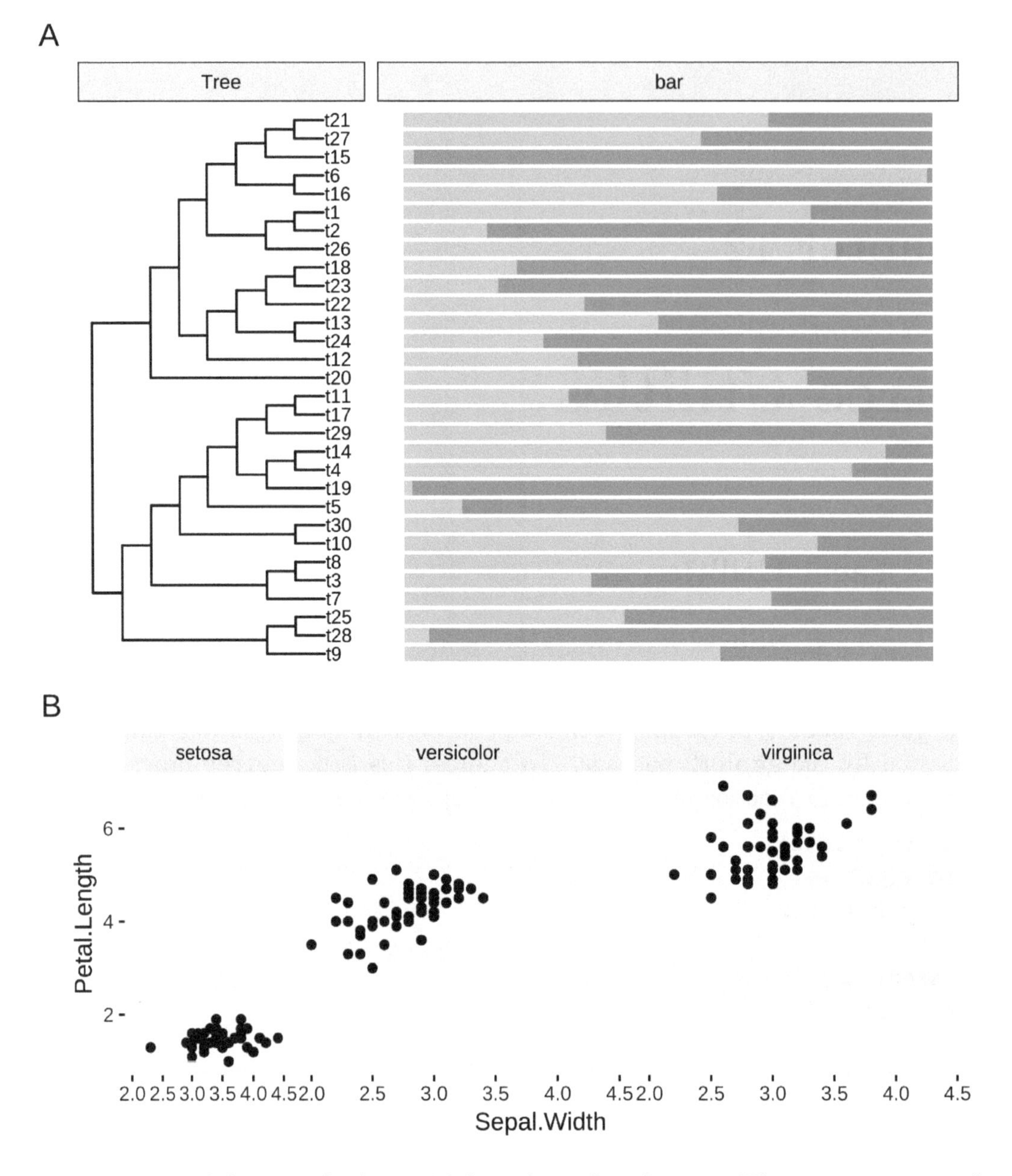

Figure 12.1: **Adjust relative widths of ggplot facets.** The `facet_widths()` function works with `ggtree` (A) as well as `ggplot` (B).

It also supports using a name vector to set the widths of specific panels. The following code will display an identical figure to Figure 12.1A.

```
facet_widths(p2, c(Tree = .5))
```

The `facet_widths()` function also works with other `ggplot` objects as demonstrated in Figure 12.1B.

```
p <- ggplot(iris, aes(Sepal.Width, Petal.Length)) +
  geom_point() + facet_grid(.~Species)
facet_widths(p, c(setosa = .5))
```

12.1.2 facet_labeller

The `facet_labeller()` function was designed to relabel selected panels (Figure 12.2), and it currently only works with `ggtree` objects (*i.e.*, `geom_facet()` outputs). A more versatile version that works with both `ggtree` and `ggplot` objects is implemented in the **ggfun** package (*i.e.*, the `facet_set()` function).

```
facet_labeller(p2, c(Tree = "phylogeny", bar = "HELLO"))
```

If you want to combine `facet_widths()` with `facet_labeller()`, you need to call `facet_labeller()` to relabel the panels before using `facet_widths()` to set the relative widths of each panel. Otherwise, it won't work since the output of `facet_widths()` is redrawn from `grid` object.

```
facet_labeller(p2, c(Tree = "phylogeny")) %>% facet_widths(c(Tree = .4))
```

12.2 Geometric Layers

Subsetting is not supported in layers defined in **ggplot2**, while it is quite useful in phylogenetic annotation since it allows us to annotate at specific node(s) (e.g., only label bootstrap values that are larger than 75).

In **ggtree**, we provide several modified versions of layers defined in **ggplot2** to support the `subset` aesthetic mapping, including:

- `geom_segment2()`
- `geom_point2()`
- `geom_text2()`
- `geom_label2()`

These layers works with both **ggtree** and **ggplot2** (Figure 12.3).

```
library(ggplot2)
library(ggtree)
data(mpg)
p <- ggplot(data = mpg, mapping = aes(x = displ, y = hwy)) +
   geom_point(mapping = aes(color = class)) +
   geom_text2(aes(label=manufacturer,
                  subset = hwy > 40 | displ > 6.5),
                  nudge_y = 1) +
   coord_cartesian(clip = "off") +
   theme_light() +
   theme(legend.position = c(.85, .75))
```

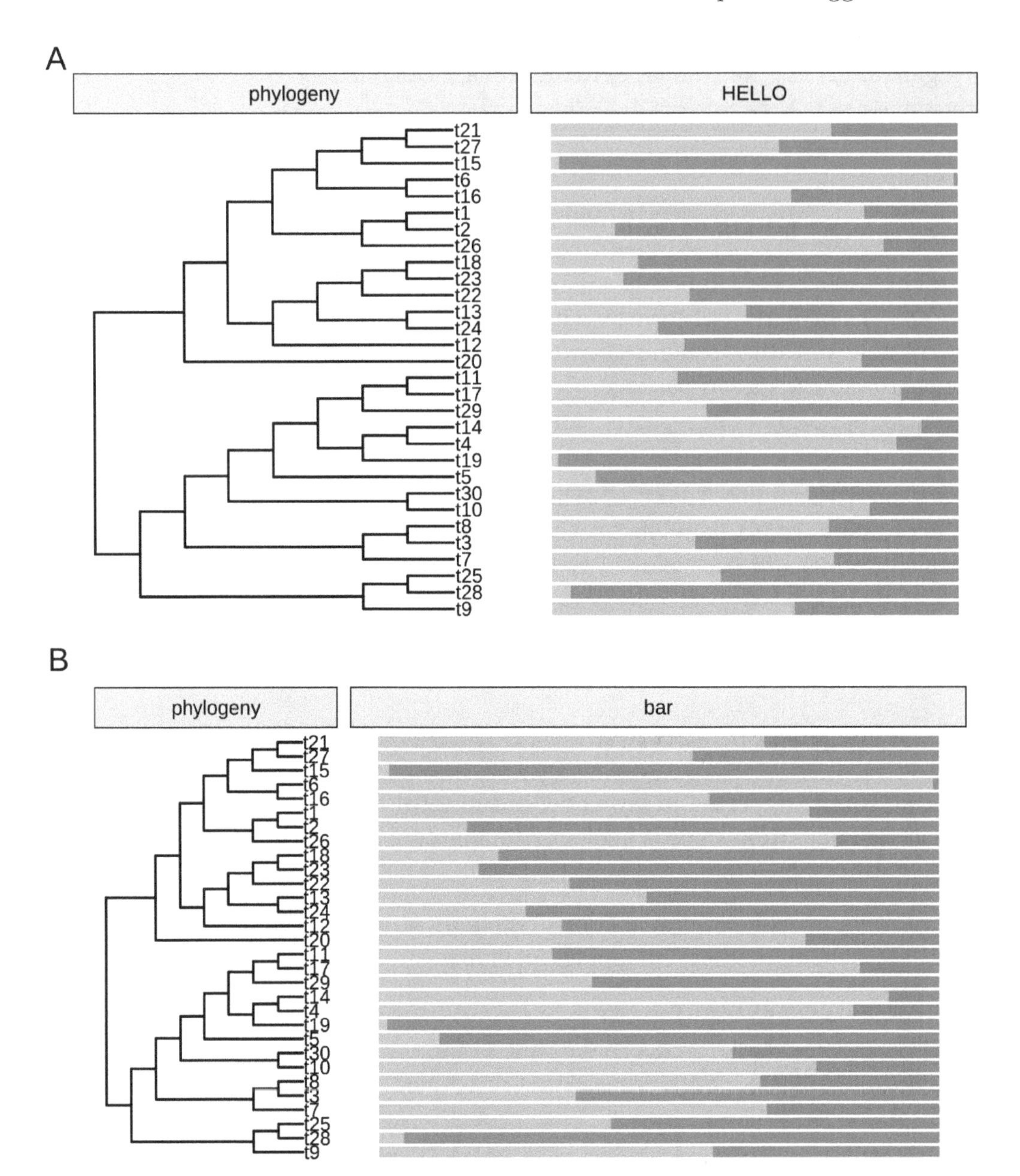

Figure 12.2: **Rename facet labels.** Rename multiple labels simultaneously (A) or only for a specific one (B) are all supported. `facet_labeller()` can combine with `facet_widths()` to rename facet label and then adjust relative widths (B).

```
p2 <- ggtree(rtree(10)) +
    geom_label2(aes(subset = node <5, label = label))

plot_list(p, p2, ncol=2, tag_levels='A')
```

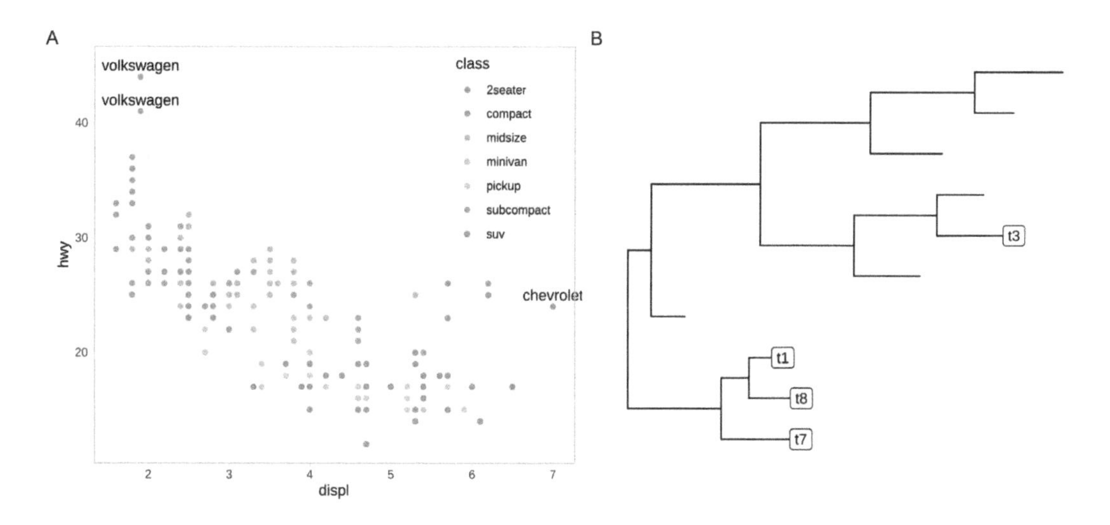

Figure 12.3: **Geometric layers that support subsetting.** These layers work with `ggplot2` (A) and `ggtree` (B).

Table 12.1: Layout transformers.

Layout	Description
layout_circular	transform rectangular layout to circular layout
layout_dendrogram	transform rectangular layout to dendrogram layout
layout_fan	transform rectangular/circular layout to fan layout
layout_rectangular	transform circular/fan layout to rectangular layout
layout_inward_circular	transform rectangular/circular layout to inward_circular layout

12.3 Layout Utilities

In session 4.2, we introduce several layouts supported by **ggtree**. The **ggtree** package also provides several layout functions that can transform from one to another. Note that not all layouts are supported (see Table 12.1 and Figure 12.4).

```
set.seed(2019)
x <- rtree(20)
p <- ggtree(x)
p + layout_dendrogram()
ggtree(x, layout = "circular") + layout_rectangular()
p + layout_circular()
p + layout_fan(angle=90)
p + layout_inward_circular(xlim=4) + geom_tiplab(hjust=1)
```

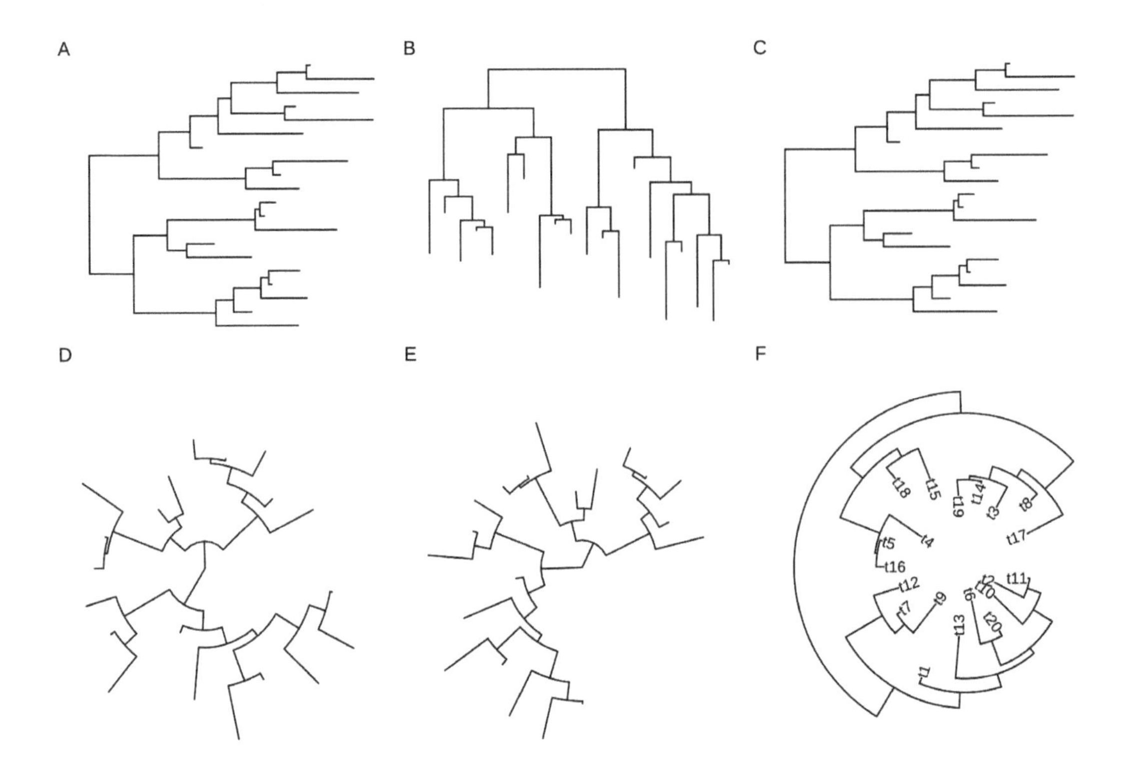

Figure 12.4: **Layout functions for transforming among different layouts**. Default rectangular layout (A); transform rectangular to dendrogram layout (B); transform circular to rectangular layout (C); transform rectangular to circular layout (D); transform rectangular to fan layout (E); transform rectangular to inward circular layout (F).

12.4 Scale Utilities

The **ggtree** package provides several scale functions to manipulate the x-axis, including the `scale_x_range()` documented in session 5.2.4, `xlim_tree()`, `xlim_expand()`, `ggexpand()`, `hexpand()` and `vexpand()`.

12.4.1 Expand x limit for a specific facet panel

Sometimes we need to set `xlim` for a specific facet panel (*e.g.*, allocate more space for long tip labels at `Tree` panel). However, the `ggplot2::xlim()` function applies to all the panels. The **ggtree** provides `xlim_expand()` to adjust `xlim` for user-specific facet panel. It accepts two parameters, `xlim`, and `panel`, and can adjust all individual panels as demonstrated in Figure 12.5A. If you only want to adjust `xlim` of the `Tree` panel, you can use `xlim_tree()` as a shortcut.

```
set.seed(2019-05-02)
x <- rtree(30)
p <- ggtree(x) + geom_tiplab()
```

```
d <- data.frame(label = x$tip.label,
                value = rnorm(30))
p2 <- p + geom_facet(panel = "Dot", data = d,
            geom = geom_point, mapping = aes(x = value))
p2 + xlim_tree(6) + xlim_expand(c(-10, 10), 'Dot')
```

The `xlim_expand()` function also works with `ggplot2::facet_grid()`. As demonstrated in Figure 12.5B, only the `xlim` of *virginica* panel was adjusted by `xlim_expand()`.

```
g <- ggplot(iris, aes(Sepal.Length, Sepal.Width)) +
    geom_point() + facet_grid(. ~ Species, scales = "free_x")
g + xlim_expand(c(0, 15), 'virginica')
```

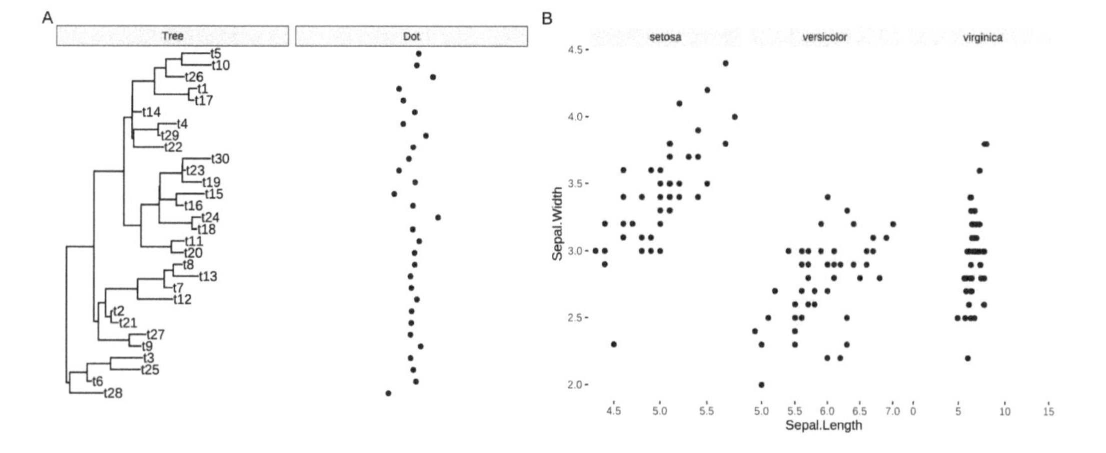

Figure 12.5: **Setting xlim for user-specific facet panels.** Using `xlim_tree()` to set the Tree panel of the `ggtree` output (A) and `xlim_expand()` to set the Dot panel of the `ggtree` output (A) and the Virginica panel of the `ggplot` output (B).

12.4.2 Expand plot limit by the ratio of plot range

The **ggplot2** package cannot automatically adjust plot limits and it is very common that long text was truncated. Users need to adjust x (y) limits manually via the `xlim()` (`ylim()`) command (see also FAQ: Tip label truncated).

The `xlim()` (`ylim()`) is a good solution to this issue. However, we can make it more simple, by expanding the plot panel by a ratio of the axis range without knowing what the exact value is.

We provide `hexpand()` function to expand x limit by specifying a fraction of the x range and it works for both directions (`direction=1` for right-hand side and `direction=-1` for left-hand side) (Figure 12.6). Another version of `vexpand()` works with similar behavior for y-axis and the `ggexpand()` function works for both x- and y-axis (Figure 11.2).

```
x$tip.label <- paste0('to make the label longer_', x$tip.label)
p1 <- ggtree(x) + geom_tiplab() + hexpand(.4)
p2 <- ggplot(iris, aes(Sepal.Width, Petal.Width)) +
    geom_point() +
    hexpand(.2, direction = -1) +
    vexpand(.2)

plot_list(p1, p2, tag_levels="A", widths=c(.6, .4))
```

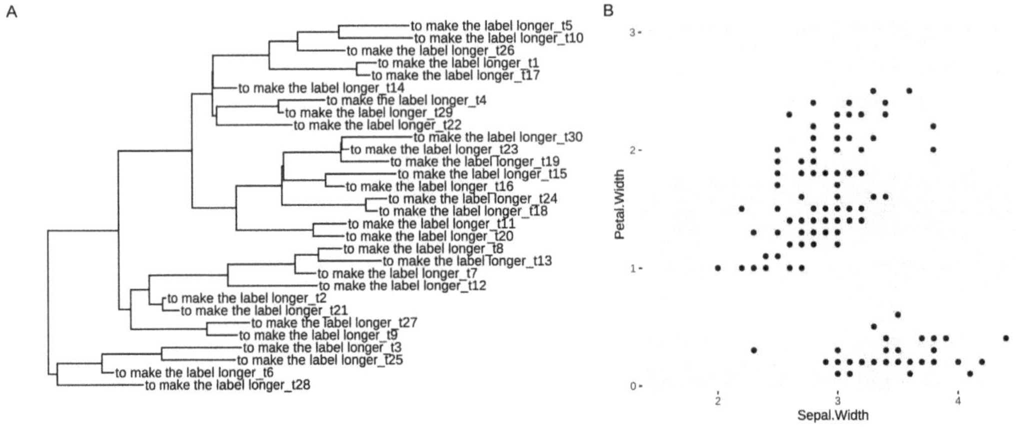

Figure 12.6: **Expanding plot limits by a fraction of the x or y range.** Expand x limit at right-hand side by default (A), and expand x limit for left-hand side when direction = -1 and expand y limit at the upper side (B).

12.5 Tree data utilities

12.5.1 Filter tree data

The **ggtree** package defined several geom layers that support subsetting tree data. However, many other geom layers that didn't provide this feature, are defined in **ggplot2** and its extensions. To allow filtering tree data with these layers, **ggtree** provides an accompanying function, `td_filter()` that returns a function that works similar to `dplyr::filter()` and can be passed to the `data` parameter in geom layers to filter `ggtree` plot data as demonstrated in Figure 12.7.

```
library(tidytree)

set.seed(1997)
tree <- rtree(50)
p <- ggtree(tree)
selected_nodes <- offspring(p, 67)$node
p + geom_text(aes(label=label),
```

```
            data=td_filter(isTip &
                           node %in% selected_nodes),
            hjust=0) +
    geom_nodepoint(aes(subset = node ==67),
                   size=5, color='blue')
```

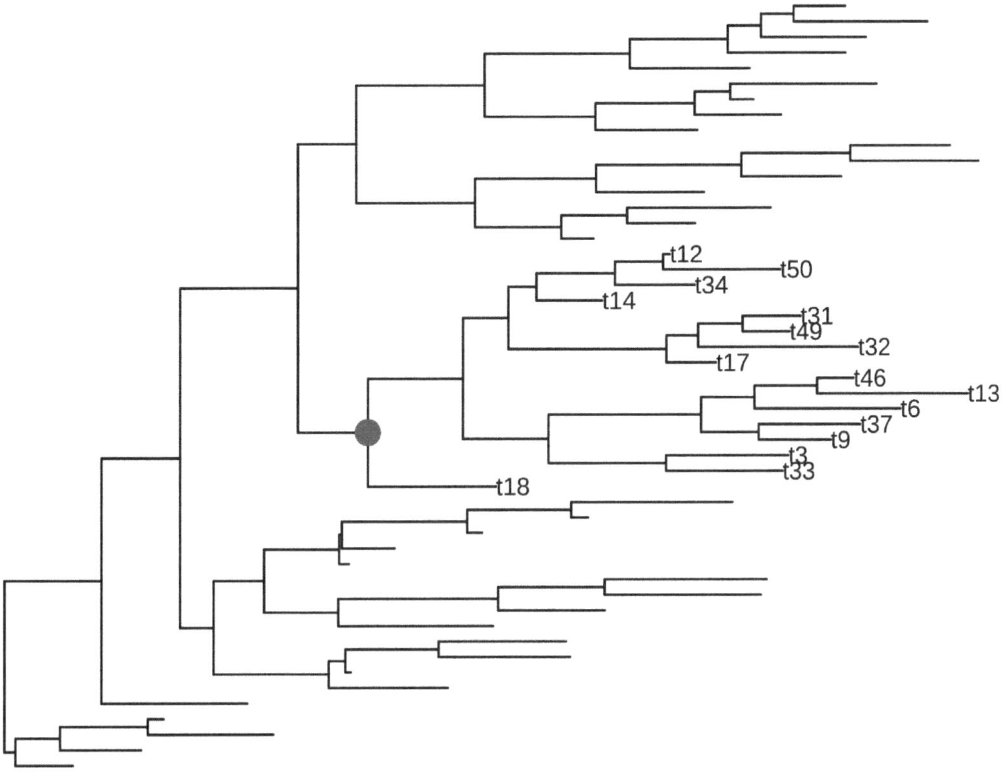

Figure 12.7: **Filtering ggtree plot data in geom layers.** Only selected tips (offspring of the node indicated by the blue circle point) were labeled.

12.5.2 Flatten list-column tree data

The `ggtree` plot data is a tidy data frame where each row represents a unique node. If multiple values are associated with a node, the data can be stored as nested data (i.e., in a list-column).

```
set.seed(1997)
tr <- rtree(5)
d <- data.frame(id=rep(tr$tip.label,2),
                value=abs(rnorm(10, 6, 2)),
                group=c(rep("A", 5),rep("B",5)))

require(tidyr)
d2  <- nest(d, value =value, group=group)
```

```
## d2 is a nested data
d2
```

```
## # A tibble: 5 x 3
##   id    value             group
##   <chr> <list>            <list>
## 1 t2    <tibble [2 x 1]> <tibble [2 x 1]>
## 2 t1    <tibble [2 x 1]> <tibble [2 x 1]>
## 3 t5    <tibble [2 x 1]> <tibble [2 x 1]>
## 4 t4    <tibble [2 x 1]> <tibble [2 x 1]>
## 5 t3    <tibble [2 x 1]> <tibble [2 x 1]>
```

Nested data is supported by the operator, `%<+%`, and can be mapped to the tree structure. If a geom layer can't directly support visualizing nested data, we need to flatten the data before applying the geom layer to display it. The **ggtree** package provides a function, `td_unnest()`, which returns a function that works similar to `tidyr::unnest()` and can be used to flatten `ggtree` plot data as demonstrated in Figure 12.8A.

All tree data utilities provide a `.f` parameter to pass a function to pre-operate the data. This creates the possibility to combine different tree data utilities as demonstrated in Figure 12.8B.

```
p <- ggtree(tr) %<+% d2
p2 <- p +
    geom_point(aes(x, y, size= value, colour=group),
            data = td_unnest(c(value, group)), alpha=.4) +
    scale_size(range=c(3,10), limits=c(3, 10))

p3 <- p +
    geom_point(aes(x, y, size= value, colour=group),
            data = td_unnest(c(value, group),
                        .f = td_filter(isTip & node==4)),
            alpha=.4) +
    scale_size(range=c(3,10), limits=c(3, 10))

plot_list(p2, p3, tag_levels = 'A')
```

12.6 Tree Utilities

12.6.1 Extract tip order

To create composite plots, users need to re-order their data manually before creating tree-associated graphs. The order of their data should be consistent with the tip order presented in the `ggtree()` plot. For this purpose, we provide the `get_taxa_name()` function to extract an ordered vector of tips based on the tree structure plotted by

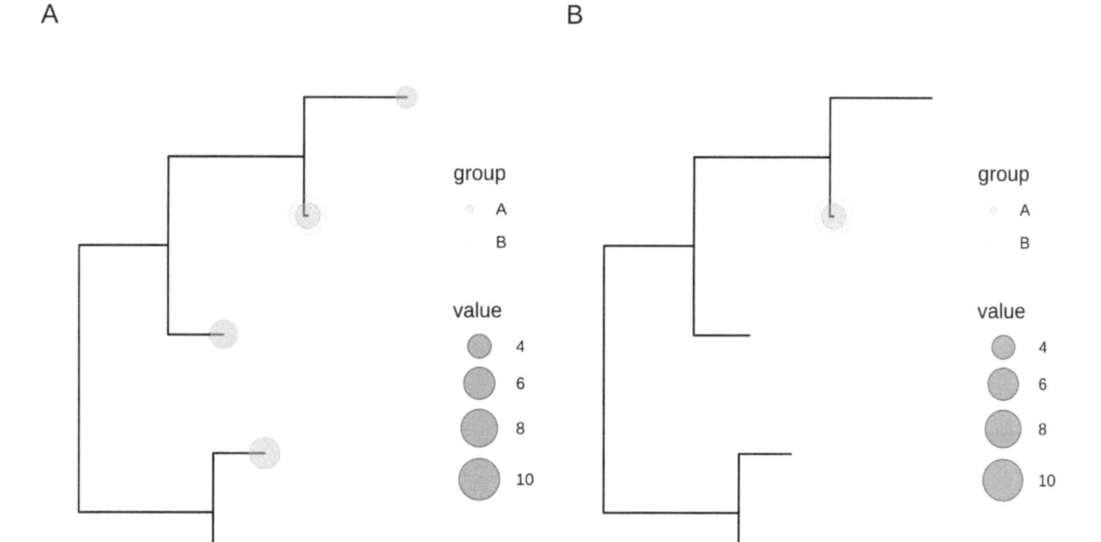

Figure 12.8: **Flattening ggtree plot data.** List-columns can be flattened by `td_unnest()` and two circle points were displayed on each tip simultaneously (A). Different tree data utilities can be combined to work together, e.g., filter data by `td_filter()`, and then flatten it by `td_unnest()`) (B).

`ggtree()`.

```
set.seed(123)
tree <- rtree(10)
p <- ggtree(tree) + geom_tiplab() +
    geom_hilight(node = 12, extendto = 2.5)

x <- paste("Taxa order:",
        paste0(get_taxa_name(p), collapse=', '))
p + labs(title=x)
```

The `get_taxa_name()` function will return a vector of ordered tip labels according to the tree structure displayed in Figure 12.9.

```
get_taxa_name(p)
```

```
##  [1] "t9"  "t8"  "t3"  "t2"  "t7"  "t10" "t1"  "t5"
##  [9] "t6"  "t4"
```

If users specify a node, the `get_taxa_name()` will extract the tip order of the selected clade (i.e., highlighted region in Figure 12.9).

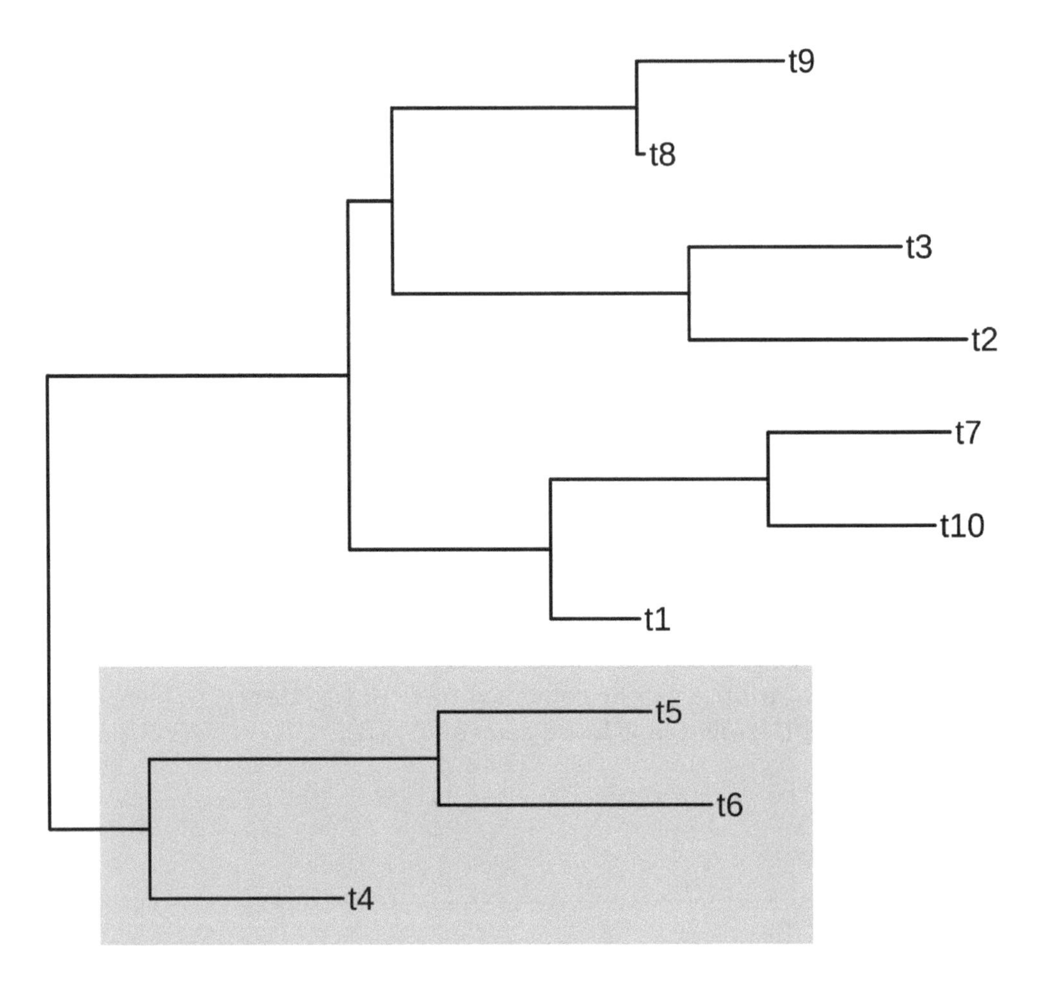

Figure 12.9: **An example tree for demonstrating `get_taxa_name()` function.**

```
get_taxa_name(p, node = 12)

## [1] "t5" "t6" "t4"
```

12.6.2 Padding taxa labels

The `label_pad()` function adds padding characters (default is ·) to taxa labels.

```
set.seed(2015-12-21)
tree <- rtree(5)
tree$tip.label[2] <- "long string for test"

d <- data.frame(label = tree$tip.label,
                newlabel = label_pad(tree$tip.label),
```

```
                newlabel2 = label_pad(tree$tip.label, pad = " "))
print(d)
```

```
##                    label           newlabel
## 1                     t1 ··················t1
## 2 long string for test long string for test
## 3                     t2 ··················t2
## 4                     t4 ··················t4
## 5                     t3 ··················t3
##                newlabel2
## 1                     t1
## 2 long string for test
## 3                     t2
## 4                     t4
## 5                     t3
```

This feature is useful if we want to align tip labels to the end as demonstrated in Figure 12.10. Note that in this case, monospace font should be used to ensure the lengths of the labels displayed in the plot are the same.

```
p <- ggtree(tree) %<+% d + xlim(NA, 5)
p1 <- p + geom_tiplab(aes(label=newlabel),
                      align=TRUE, family='mono',
                      linetype = "dotted", linesize = .7)
p2 <- p + geom_tiplab(aes(label=newlabel2),
                      align=TRUE, family='mono',
                      linetype = NULL, offset=-.5) + xlim(NA, 5)
plot_list(p1, p2, ncol=2, tag_levels = "A")
```

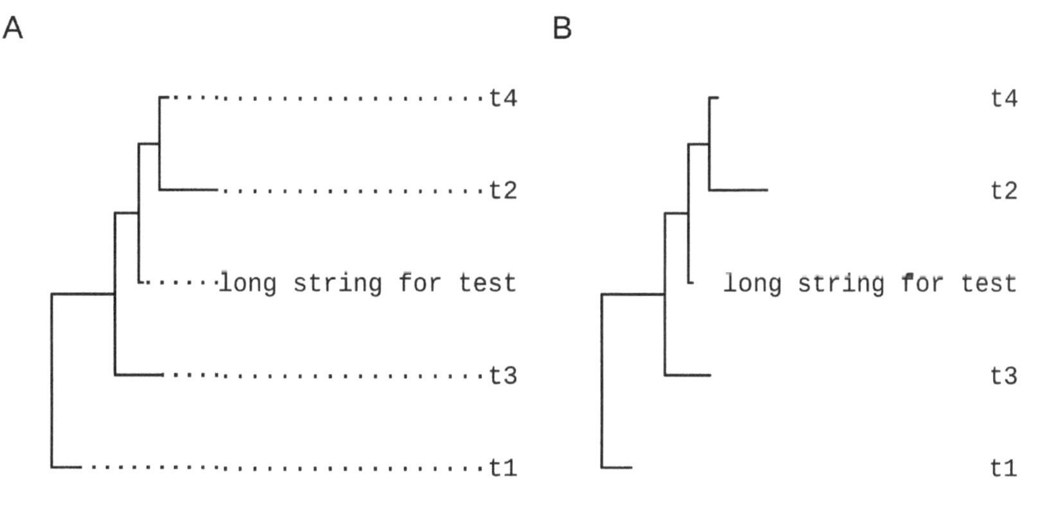

Figure 12.10: **Align tip label to the end.** With a dotted line (A) and without a dotted line (B).

12.7 Interactive ggtree Annotation

The **ggtree** package supports interactive tree annotation or manipulation by implementing an `identify()` method. Users can click on a node to highlight a clade, to label or rotate it, *etc*. Users can also use the **plotly** package to convert a `ggtree` object to a `plotly` object to quickly create an interactive phylogenetic tree.

Video of using `identify()` to interactively manipulate a phylogenetic tree can be found on Youtube and Youku:

- Highlighting clades: Youtube and Youku.
- Labelling clades: Youtube and Youku.
- Rotating clades: Youtube and Youku.

Chapter 13

Gallery of Reproducible Examples

13.1 Visualizing pairwise nucleotide sequence distance with a phylogenetic tree

This example reproduces figure 1 of (Chen et al., 2017). It extracts accession numbers from tip labels of the HPV58 tree and calculates pairwise nucleotide sequence distances. The distance matrix is visualized as dot and line plots. This example demonstrates the ability to add multiple layers to a specific panel. As illustrated in Figure 13.1, the `geom_facet()` function displays sequence distances as a dot plot and then adds a layer of line plot to the same panel, *i.e.*, sequence distance. In addition, the tree in `geom_facet()` can be fully annotated with multiple layers (clade labels, bootstrap support values, *etc.*). The source code is modified from the supplemental file of (Yu et al., 2018).

```
library(TDbook)
library(tibble)
library(tidyr)
library(Biostrings)
library(treeio)
library(ggplot2)
library(ggtree)

# loaded from TDbook package
tree <- tree_HPV58

clade <- c(A3 = 92, A1 = 94, A2 = 108, B1 = 156,
            B2 = 159, C = 163, D1 = 173, D2 = 176)
tree <- groupClade(tree, clade)
cols <- c(A1 = "#EC762F", A2 = "#CA6629", A3 = "#894418", B1 = "#0923FA",
         B2 = "#020D87", C = "#000000", D1 = "#9ACD32",D2 = "#08630A")
```

```
## visualize the tree with tip labels and tree scale
p <- ggtree(tree, aes(color = group), ladderize = FALSE) %>%
    rotate(rootnode(tree)) +
    geom_tiplab(aes(label = paste0("italic('", label, "')")),
                parse = TRUE, size = 2.5) +
    geom_treescale(x = 0, y = 1, width = 0.002) +
    scale_color_manual(values = c(cols, "black"),
                na.value = "black", name = "Lineage",
                breaks = c("A1", "A2", "A3", "B1", "B2", "C", "D1", "D2")) +
    guides(color = guide_legend(override.aes = list(size = 5, shape = 15))) +
    theme_tree2(legend.position = c(.1, .88))
## Optional
## add labels for monophyletic (A, C and D) and paraphyletic (B) groups
dat <- tibble(node = c(94, 108, 131, 92, 156, 159, 163, 173, 176,172),
              name = c("A1", "A2", "A3", "A", "B1",
                       "B2", "C", "D1", "D2", "D"),
              offset = c(0.003, 0.003, 0.003, 0.00315, 0.003,
                         0.003, 0.0031, 0.003, 0.003, 0.00315),
              offset.text = c(-.001, -.001, -.001, 0.0002, -.001,
                         -.001, 0.0002, -.001, -.001, 0.0002),
              barsize = c(1.2, 1.2, 1.2, 2, 1.2, 1.2, 3.2, 1.2, 1.2, 2),
              extend = list(c(0, 0.5), 0.5, c(0.5, 0), 0, c(0, 0.5),
                         c(0.5, 0), 0, c(0, 0.5), c(0.5, 0), 0)
          ) %>%
       dplyr::group_split(barsize)

p <- p +
     geom_cladelab(
         data = dat[[1]],
         mapping = aes(
             node = node,
             label = name,
             color = group,
             offset = offset,
             offset.text = offset.text,
             extend = extend
         ),
         barsize = 1.2,
         fontface = 3,
         align = TRUE
     ) +
     geom_cladelab(
         data = dat[[2]],
         mapping = aes(
             node = node,
             label = name,
             offset = offset,
             offset.text =offset.text,
             extend = extend
         ),
```

```
        barcolor = "darkgrey",
        textcolor = "darkgrey",
        barsize = 2,
        fontsize = 5,
        fontface = 3,
        align = TRUE
    ) +
    geom_cladelab(
        data = dat[[3]],
        mapping = aes(
            node = node,
            label = name,
            offset = offset,
            offset.text = offset.text,
            extend = extend
        ),
        barcolor = "darkgrey",
        textcolor = "darkgrey",
        barsize = 3.2,
        fontsize = 5,
        fontface = 3,
        align = TRUE
    ) +
    geom_strip(65, 71, "italic(B)", color = "darkgrey",
               offset = 0.00315, align = TRUE, offset.text = 0.0002,
               barsize = 2, fontsize = 5, parse = TRUE)

## Optional
## display support values
p <- p + geom_nodelab(aes(subset = (node == 92), label = "*"),
                      color = "black", nudge_x = -.001, nudge_y = 1) +
    geom_nodelab(aes(subset = (node == 155), label = "*"),
                      color = "black", nudge_x = -.0003, nudge_y = -1) +
    geom_nodelab(aes(subset = (node == 158), label = "95/92/1.00"),
                      color = "black", nudge_x = -0.0001,
                      nudge_y = -1, hjust = 1) +
    geom_nodelab(aes(subset = (node == 162), label = "98/97/1.00"),
                      color = "black", nudge_x = -0.0001,
                      nudge_y = -1, hjust = 1) +
    geom_nodelab(aes(subset = (node == 172), label = "*"),
                      color = "black", nudge_x = -.0003, nudge_y = -1)

## extract accession numbers from tip labels
tl <- tree$tip.label
acc <- sub("\\w+\\|", "", tl)
names(tl) <- acc

## read sequences from GenBank directly into R
## and convert the object to DNAStringSet
```

```
tipseq <- ape::read.GenBank(acc) %>% as.character %>%
    lapply(., paste0, collapse = "") %>% unlist %>%
    DNAStringSet
## align the sequences using muscle
tipseq_aln <- muscle::muscle(tipseq)
tipseq_aln <- DNAStringSet(tipseq_aln)

## calculate pairwise hamming distances among sequences
tipseq_dist <- stringDist(tipseq_aln, method = "hamming")

## calculate the percentage of differences
tipseq_d <- as.matrix(tipseq_dist) / width(tipseq_aln[1]) * 100

## convert the matrix to a tidy data frame for facet_plot
dd <- as_tibble(tipseq_d)
dd$seq1 <- rownames(tipseq_d)
td <- gather(dd,seq2, dist, -seq1)
td$seq1 <- tl[td$seq1]
td$seq2 <- tl[td$seq2]

g <- p$data$group
names(g) <- p$data$label
td$clade <- g[td$seq2]

## visualize the sequence differences using dot plot and line plot
## and align the sequence difference plot to the tree using facet_plot
p2 <- p + geom_facet(panel = "Sequence Distance",
            data = td, geom = geom_point, alpha = .6,
            mapping = aes(x = dist, color = clade, shape = clade)) +
    geom_facet(panel = "Sequence Distance",
            data = td, geom = geom_path, alpha = .6,
            mapping=aes(x = dist, group = seq2, color = clade)) +
    scale_shape_manual(values = 1:8, guide = FALSE)

print(p2)
```

13.2 Displaying Different Symbolic Points for Bootstrap Values.

We can cut the bootstrap values into several intervals, *e.g.*, to indicate whether the clade is of high, moderate, or low support. Then we can use these intervals as categorical variables to set different colors or shapes of symbolic points to indicate the bootstrap values belong to which category (Figure 13.2).

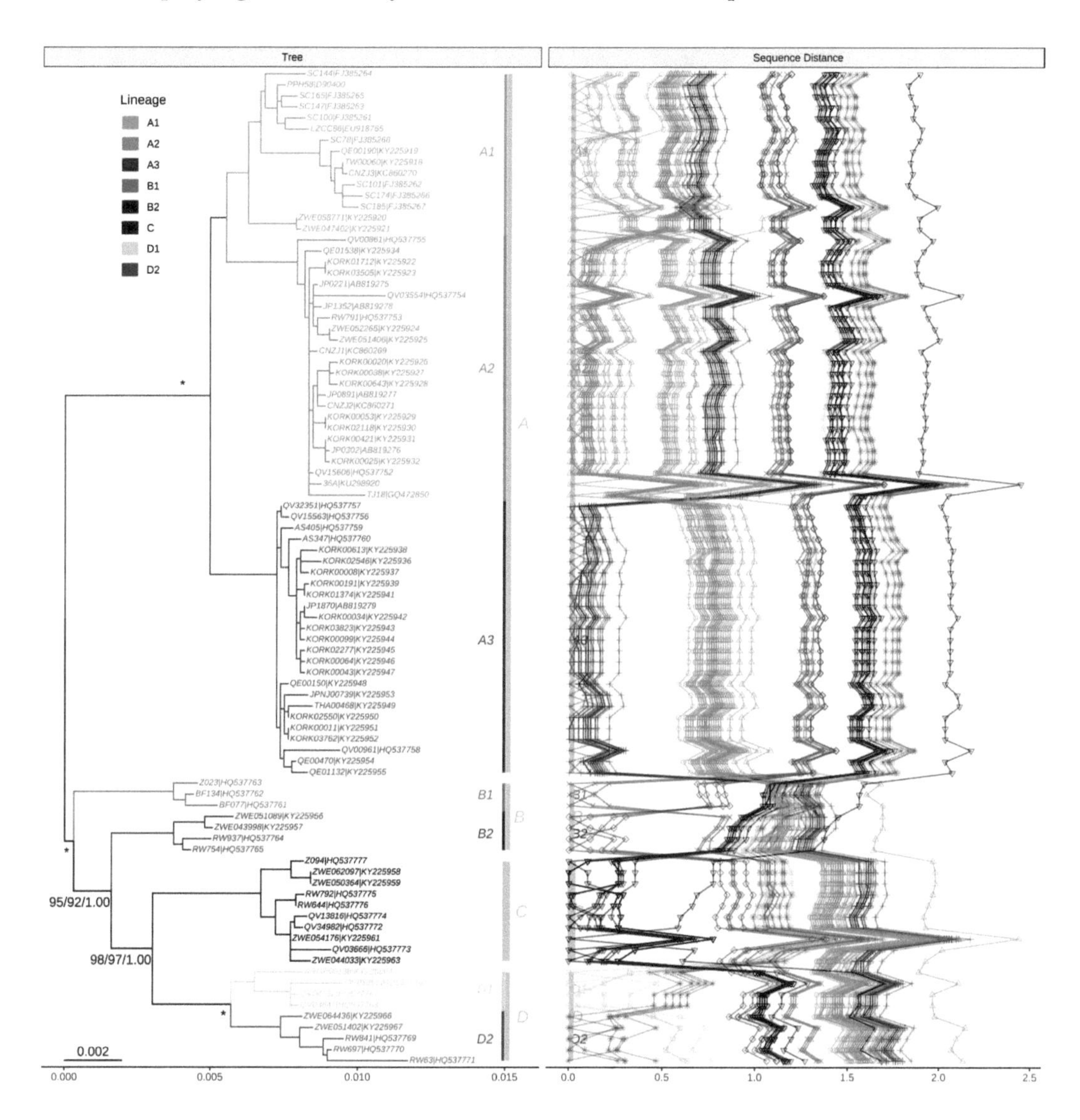

Figure 13.1: **Phylogeny of HPV58 complete genomes with dot and line plots of pairwise nucleotide sequence distances**.

```
library(treeio)
library(ggplot2)
library(ggtree)
library(TDbook)

tree <- read.newick(text=text_RMI_tree, node.label = "support")
root <- rootnode(tree)
ggtree(tree, color="black", size=1.5, linetype=1,  right=TRUE) +
    geom_tiplab(size=4.5, hjust = -0.060, fontface="bold") + xlim(0, 0.09) +
    geom_point2(aes(subset=!isTip & node != root,
                    fill=cut(support, c(0, 700, 900, 1000))),
```

```
                  shape=21, size=4) +
    theme_tree(legend.position=c(0.2, 0.2)) +
    scale_fill_manual(values=c("white", "grey", "black"), guide='legend',
                  name='Bootstrap Percentage(BP)',
                  breaks=c('(900,1e+03]', '(700,900]', '(0,700]'),
                  labels=expression(BP>=90,70 <= BP * " < 90", BP < 70))
```

13.3 Highlighting Different Groups

This example reproduces Figure 1 of (Larsen et al., 2019). It used `groupOTU()` to add grouping information of chicken CTLDcps. The branch line type and color are defined based on this grouping information. Two groups of CTLDcps are highlighted in different background colors using `geom_hilight` (red for Group II and green for Group V). The avian-specific expansion of Group V with the subgroups of A and B- are labeled using `geom_cladelab` (Figure 13.3).

```
library(TDbook)
mytree <- tree_treenwk_30.4.19

# Define nodes for coloring later on
tiplab <- mytree$tip.label
cls <- tiplab[grep("^ch", tiplab)]
labeltree <- groupOTU(mytree, cls)

p <- ggtree(labeltree, aes(color=group, linetype=group),
            layout="circular") +
    scale_color_manual(values = c("#efad29", "#63bbd4")) +
    geom_nodepoint(color="black", size=0.1) +
    geom_tiplab(size=2, color="black")

p2 <- flip(p, 136, 110) %>%
    flip(141, 145) %>%
    rotate(141) %>%
    rotate(142) %>%
    rotate(160) %>%
    rotate(164) %>%
    rotate(131)

### Group V and II coloring
dat <- data.frame(
          node = c(110, 88, 156,136),
          fill = c("#229f8a", "#229f8a", "#229f8a", "#f9311f")
      )
p3 <- p2 +
      geom_hilight(
```

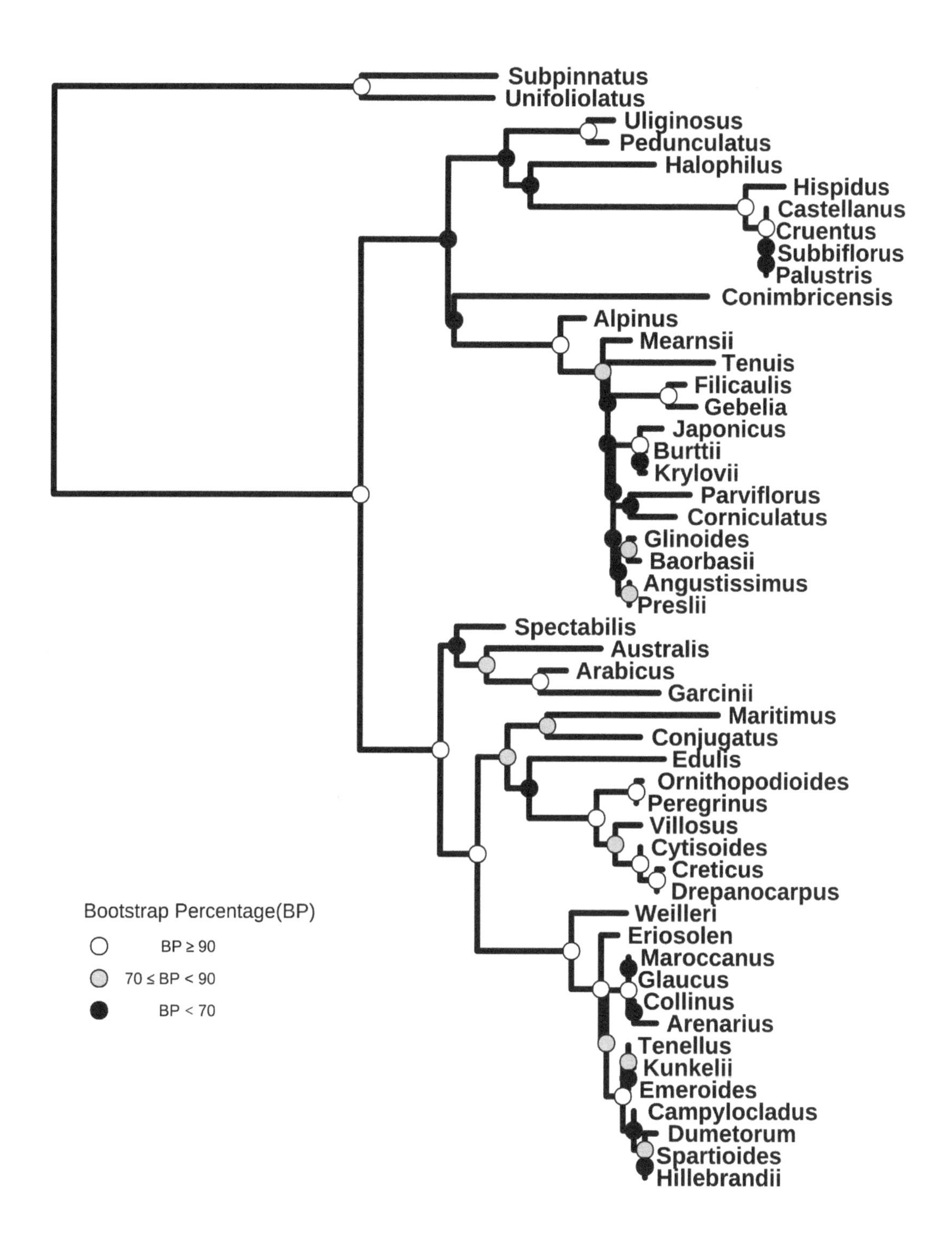

Figure 13.2: **Partitioning bootstrap values**. Bootstrap values were divided into three categories and this information was used to color circle points.

```
        data = dat,
        mapping = aes(
            node = node,
            fill = I(fill)
        ),
        alpha = 0.2,
        extendto = 1.4
    )

### Putting on a label on the avian specific expansion
p4 <- p3 +
    geom_cladelab(
        node = 113,
        label = "Avian-specific expansion",
        align = TRUE,
        angle = -35,
        offset.text = 0.05,
        hjust = "center",
        fontsize = 2,
        offset = .2,
        barsize = .2
    )

### Adding the bootstrap values with subset used to remove all
### bootstraps < 50
p5 <- p4 +
    geom_nodelab(
        mapping = aes(
            x = branch,
            label = label,
            subset = !is.na(as.numeric(label)) & as.numeric(label)
            > 50
        ),
        size = 2,
        color = "black",
        nudge_y = 0.6
    )

### Putting labels on the subgroups
p6 <- p5 +
    geom_cladelab(
        data = data.frame(
            node = c(114, 121),
            name = c("Subgroup A", "Subgroup B")
```

```
        ),
        mapping = aes(
            node = node,
            label = name
        ),
        align = TRUE,
        offset = .05,
        offset.text = .03,
        hjust = "center",
        barsize = .2,
        fontsize = 2,
        angle = "auto",
        horizontal = FALSE
    ) +
    theme(
        legend.position = "none",
        plot.margin = grid::unit(c(-15, -15, -15, -15), "mm")
    )
print(p6)
```

13.4 Phylogenetic Tree with Genome Locus Structure

The `geom_motif()` is defined in **ggtree** and it is a wrapper layer of the `gggenes::geom_gene_arrow()`. The `geom_motif()` can automatically adjust genomic alignment by selective gene (via the `on` parameter) and can label genes via the `label` parameter. In the following example, we use `example_genes` dataset provided by **gggenes**. As the dataset only provides genomic coordination of a set of genes, a phylogeny for the genomes needs to be constructed first. We calculate Jaccard similarity based on the ratio of overlapping genes among genomes and correspondingly determine genome distance. The BioNJ algorithm was applied to construct the tree. Then we can use `geom_facet()` to visualize the tree with the genomic structures (Figure 13.4).

```
library(dplyr)
library(ggplot2)
library(gggenes)
library(ggtree)

get_genes <- function(data, genome) {
    filter(data, molecule == genome) %>% pull(gene)
}

g <- unique(example_genes[,1])
n <- length(g)
```

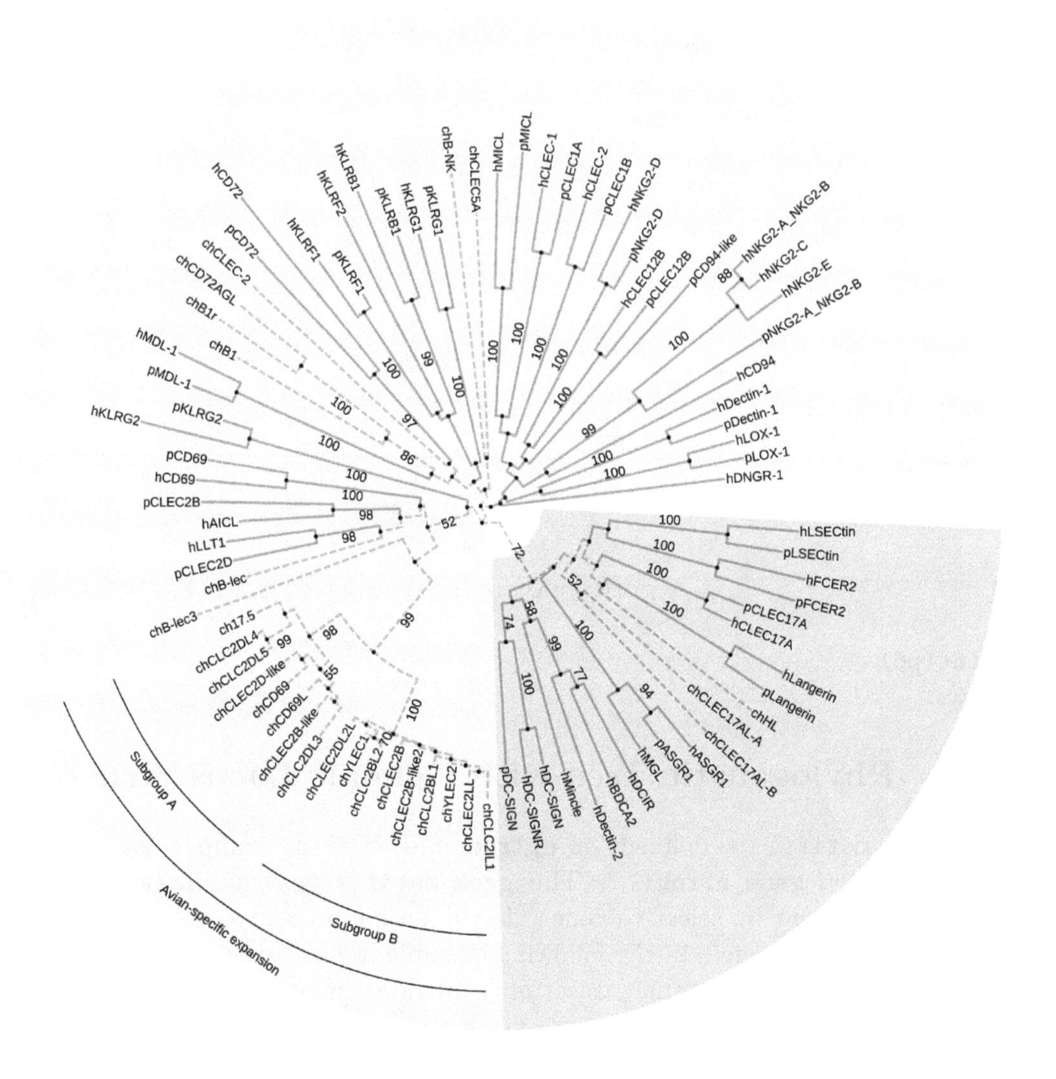

Figure 13.3: **Phylogenetic tree of CTLDcps**. Using different background colors, line types and colors, and clade labels to distinguish groups.

```
d <- matrix(nrow = n, ncol = n)
rownames(d) <- colnames(d) <- g
genes <- lapply(g, get_genes, data = example_genes)

for (i in 1:n) {
    for (j in 1:i) {
        jaccard_sim <- length(intersect(genes[[i]], genes[[j]])) /
                       length(union(genes[[i]], genes[[j]]))
        d[j, i] <- d[i, j] <- 1 - jaccard_sim
    }
```

```
}

tree <- ape::bionj(d)

p <- ggtree(tree, branch.length='none') +
    geom_tiplab() + xlim_tree(5.5) +
    geom_facet(mapping = aes(xmin = start, xmax = end, fill = gene),
               data = example_genes, geom = geom_motif,
               panel = 'Alignment',
               on = 'genE', label = 'gene', align = 'left') +
    scale_fill_brewer(palette = "Set3") +
    scale_x_continuous(expand=c(0,0)) +
    theme(strip.text=element_blank(),
        panel.spacing=unit(0, 'cm'))

facet_widths(p, widths=c(1,2))
```

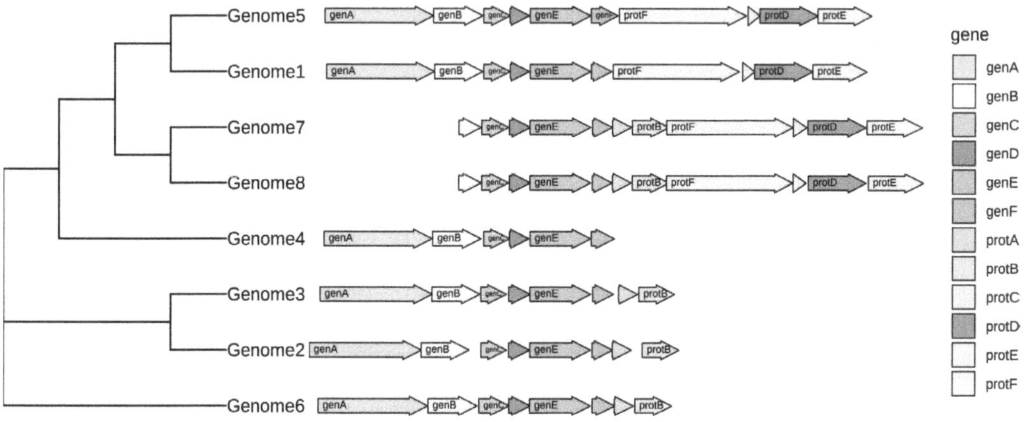

Figure 13.4: **Genomic features with a phylogenetic tree.**

Appendix A

Frequently Asked Questions

The ggtree mailing-list[1] is a great place to get help, once you have created a reproducible example that illustrates your problem.

A.1 Installation

The **ggtree** is released within the Bioconductor project; you need to use **BiocManager** to install it.

```
## you need to install BiocManager before using it
## install.packages("BiocManager")
library(BiocManager)
install("ggtree")
```

Bioconductor release is adhered to a specific R version. Please make sure you are using the latest version of R if you want to install the latest release of Bioconductor packages, including **ggtree**. Beware that bugs will only be fixed in the current release and develop branches. If you find a bug, please follow the guide[2] to report it.

To make it easy to install and load multiple core packages in a single step, we created a meta-package, **treedataverse**. Users can install the package via the following command:

```
BiocManager::install("YuLab-SMU/treedataverse")
```

Once it is installed, loading the package will also load the core **treedataverse** packages, including **tidytree**, **treeio**, **ggtree**, and **ggtreeExtra**.

[1]https://groups.google.com/forum/?#!forum/bioc-ggtree
[2]https://guangchuangyu.github.io/2016/07/how-to-bug-author/

A.2 Basic R Related

A.2.1 Use your local file

If you are new to R and want to use **ggtree** for tree visualization, please do learn some basic R and **ggplot2**.

A very common issue is that users copy and paste commands without looking at the function's behavior. The `system.file()` function was used in some of our examples to find files packed in the packages.

```
system.file                  package:base                  R Documentation

Find Names of R System Files

Description:

     Finds the full file names of files in packages etc.

Usage:

     system.file(..., package = "base", lib.loc = NULL,
                 mustWork = FALSE)
```

For users who want to use their files, please just use relative or absolute file path (*e.g.*, `file = "your/folder/filename"`).

A.3 Aesthetic mapping

A.3.1 Inherit aesthetic mapping

```
ggtree(rtree(30)) + geom_point()
```

For example, we can add symbolic points to nodes with `geom_point()` directly. The magic here is we don't need to map the `x` and `y` position of the points by providing `aes(x, y)` to `geom_point()` since it was already mapped by the `ggtree()` function and it serves as a global mapping for all layers.

But what if we provide a dataset in a layer and the dataset doesn't contain columns of `x` and/or `y`, the layer function also tries to map `x` and `y` and also others if you map them in the `ggtree()` function. As these variables are not available in your dataset, you will get the following error:

```
Error in eval(expr, envir, enclos) : object 'x' not found
```

This can be fixed by using the parameter `inherit.aes=FALSE` which will disable inheriting mapping from the `ggtree()` function.

A.3.2 Never use $ in aesthetic mapping

Never do this[3] and please refer to the explanation in the ggplot2 book 2ed (Wickham, 2016):

> Never refer to a variable with `$` (e.g., `diamonds$carat`) in `aes()`. This breaks containment so that the plot no longer contains everything it needs and causes problems if ggplot2 changes the order of the rows, as it does when facetting.

A.4 Text and Label

A.4.1 Tip label truncated

The reason for this issue is that **ggplot2** can't auto-adjust `xlim` based on added text[4].

```
library(ggtree)
## example tree from https://support.bioconductor.org/p/72398/
tree <- read.tree(text= paste("(Organism1.006G249400.1:0.03977,",
    "(Organism2.022118m:0.01337,(Organism3.J34265.1:0.00284,",
    "Organism4.G02633.1:0.00468)0.51:0.0104):0.02469);"))
p <- ggtree(tree) + geom_tiplab()
```

In this example, the tip labels displayed in Figure A.1A are truncated. This is because the units are in two different spaces (data and pixel). Users can use `xlim` to allocate more spaces for tip labels (Figure A.1B).

```
p + xlim(0, 0.08)
```

Another solution is to set `clip = "off"` to allow drawing outside of the plot panel. We may also need to set `plot.margin` to allocate more spaces for margin (Figure A.1C).

```
p + coord_cartesian(clip = 'off') +
  theme_tree2(plot.margin=margin(6, 120, 6, 6))
```

The third solution is to use `hexpand()` as demonstrated in session 12.4.

For rectangular/dendrogram layout trees, users can display tip labels as y-axis labels. In this case, no matter how long the labels are, they will not be truncated (see Figure 4.8C).

---

[3] https://groups.google.com/d/msg/bioc-ggtree/hViM6vRZF94/MsZT8qRgBwAJ and https://github.com/GuangchuangYu/ggtree/issues/106

[4] https://twitter.com/hadleywickham/status/600280284869697538

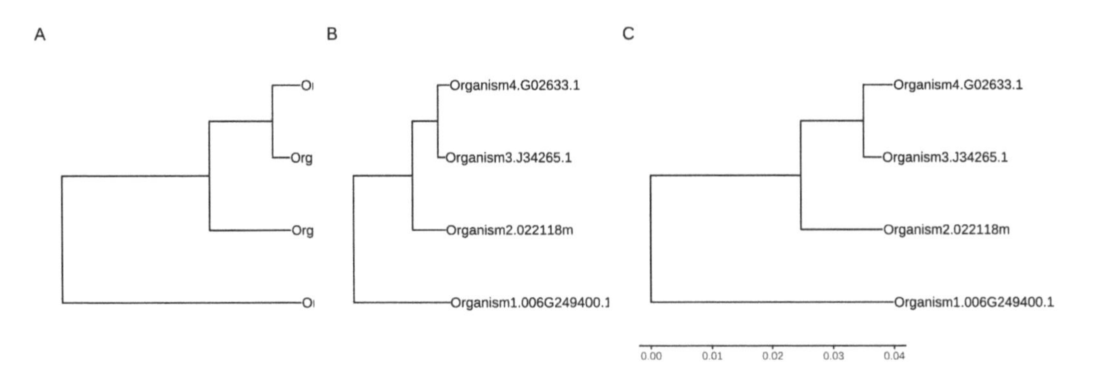

Figure A.1: **Allocating more spaces for truncated tip labels.** Long tip labels may be truncated (A). One solution is to allocate more spaces for plot panel (B), and another solution is to allow plotting labels outside the plot panel (C).

A.4.2 Modify (tip) labels

If you want to modify tip labels of the tree, you can use `treeio::rename_taxa()` to rename a `phylo` or `treedata` object.

```
tree <- read.tree(text = "((A, B), (C, D));")
d <- data.frame(label = LETTERS[1:4],
                label2 = c("sunflower", "tree", "snail", "mushroom"))

## rename_taxa use 1st column as key and 2nd column as value by default
## rename_taxa(tree, d)
rename_taxa(tree, d, label, label2) %>% write.tree

## [1] "((sunflower,tree),(snail,mushroom));"
```

If the input tree object is a `treedata` instance, you can use `write.beast()` to export the tree with associated data to a BEAST compatible NEXUS file (see Chapter 3).

Renaming phylogeny tip labels seems not to be a good idea, since it may introduce problems when mapping the original sequence alignment to the tree. Personally, I recommend storing the new labels as a tip annotation in `treedata` object.

```
tree2 <- full_join(tree, d, by = "label")
tree2

## 'treedata' S4 object'.
##
## ...@ phylo:
##
## Phylogenetic tree with 4 tips and 3 internal nodes.
##
## Tip labels:
```

```
##   A, B, C, D
##
## Rooted; no branch lengths.
##
## with the following features available:
##   'label2'.
```

If you just want to show different or additional information when plotting the tree, you don't need to modify tip labels. This could be easily done via the `%<+%` operator to attach the modified version of the labels and then use the `geom_tiplab()` layer to display the modified version (Figure A.2).

```
p <- ggtree(tree) + xlim(NA, 3)
p1 <- p + geom_tiplab()

## the following command will produce an identical figure of p2
## ggtree(tree2) + geom_tiplab(aes(label = label2))
p2 <- p %<+% d + geom_tiplab(aes(label=label2))
plot_list(p1, p2, ncol=2, tag_levels = "A")
```

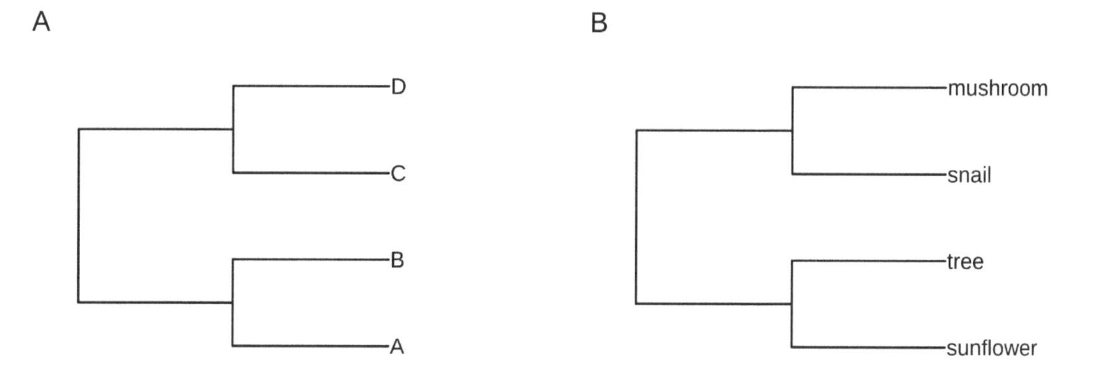

Figure A.2: **Alternative tip labels.** Original tip labels (A) and a modified version (B).

A.4.3 Formatting (tip) labels

If you want to format labels, you need to set `parse=TRUE` in the `geom_text()`/`geom_tiplab()`/`geom_nodelab()` and the `label` should be a string that can be parsed into expression and displayed as described in `?plotmath`. Users can use the **latex2exp** package to convert LaTeX math formulas to R's plotmath expressions, or use the **ggtext** package to render Markdown or HTML.

For example, the tip labels contain several parts (*e.g.*, genus, species, and geo), we can differentiate these pieces of information with different formats (Figure A.3A).

```
tree <- read.tree(text = "((a,(b,c)),d);")
genus <- c("Gorilla", "Pan", "Homo", "Pongo")
species <- c("gorilla", "spp.", "sapiens", "pygmaeus")
geo <- c("Africa", "Africa", "World", "Asia")
d <- data.frame(label = tree$tip.label, genus = genus,
                species = species, geo = geo)

library(glue)
d2 <- dplyr::mutate(d,
  lab = glue("italic({genus})~bolditalic({species})~({geo})"),
  color = c("#E495A5", "#ABB065", "#39BEB1", "#ACA4E2"),
  name = glue("<i style='color:{color}'>{genus} **{species}**</i> ({geo})")
)

p1 <- ggtree(tree) %<+% d2 + xlim(NA, 6) +
    geom_tiplab(aes(label=lab), parse=T)
```

Using Markdown or HTML to format text may be easier, and this is supported via the **ggtext** package (Figure A.3B).

```
library(ggtext)

p2 <- ggtree(tree) %<+% d2 +
  geom_richtext(data=td_filter(isTip),
                aes(label=name), label.color=NA) +
  hexpand(.3)

plot_list(p1, p2, ncol=2, tag_levels = 'A')
```

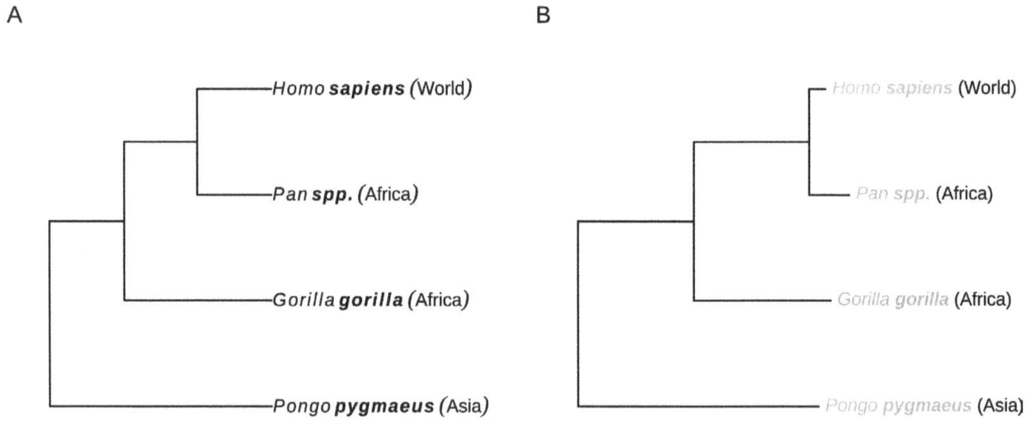

Figure A.3: **Formatting labels.** Formatting specific tip labels using `plotmath` expression (A), and Markdown/HTML (B).

A.4.4 Avoid overlapping text labels

Users can use the **ggrepel** package to repel overlapping text labels (Figure A.4).

```
library(ggrepel)
library(ggtree)
raxml_file <- system.file("extdata/RAxML",
                          "RAxML_bipartitionsBranchLabels.H3", package="treeio")
raxml <- read.raxml(raxml_file)
ggtree(raxml) + geom_label_repel(aes(label=bootstrap, fill=bootstrap)) +
  theme(legend.position = c(.1, .8)) + scale_fill_viridis_c()
```

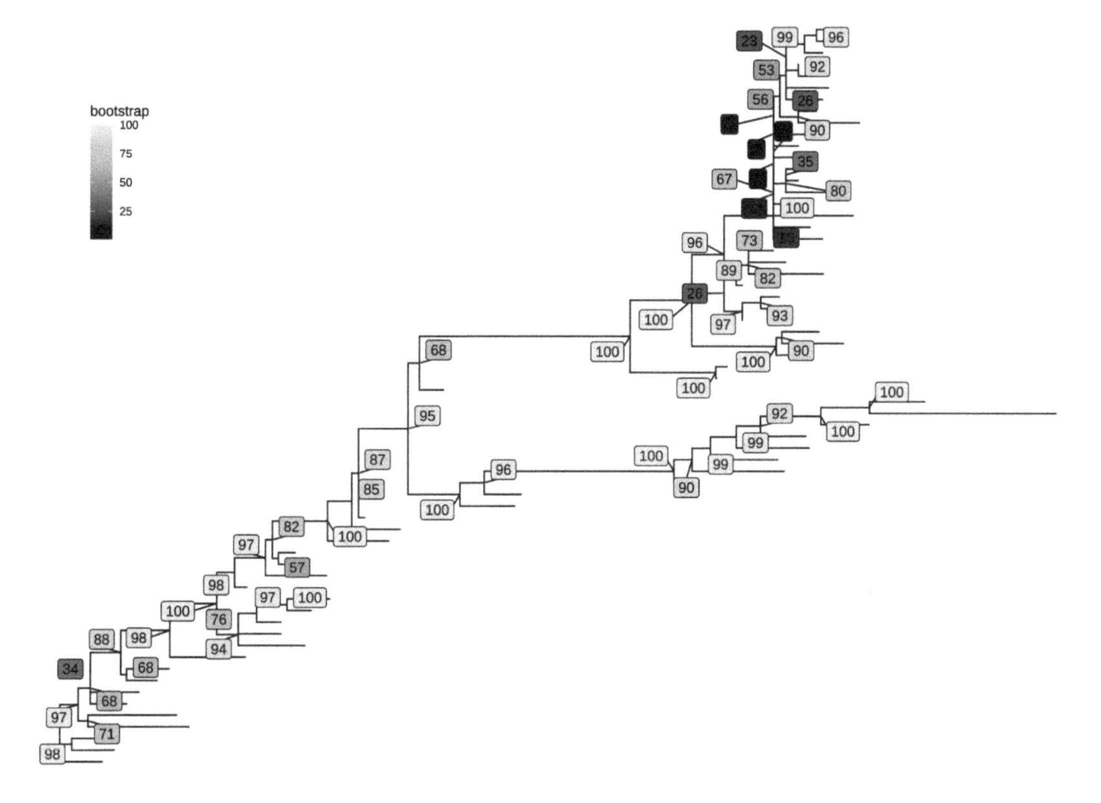

Figure A.4: **Repel labels.** Repel labels to avoid overlapping.

A.4.5 Bootstrap values from Newick format

It is quite common to store *bootstrap* value as node label in the Newick format as in Figure A.5. Visualizing node label is easy using `geom_text2(aes(subset = !isTip, label=label))`.

If you want to only display a subset of *bootstrap* (*e.g.*, bootstrap > 80), you can't simply use `geom_text2(subset= (label > 80), label=label)` (or `geom_label2`) since `label` is a character vector, which contains node label (bootstrap value) and tip label (taxa name). `geom_text2(subset=(as.numeric(label) > 80), label=label)` won't work either, since `NAs` were introduced by coercion. We need to convert `NA`s to logical `FALSE`. This can be done by the following code:

```
nwk <- system.file("extdata/RAxML","RAxML_bipartitions.H3", package='treeio')
tr <- read.tree(nwk)
ggtree(tr) + geom_label2(aes(label=label,
      subset = !is.na(as.numeric(label)) & as.numeric(label) > 80))
```

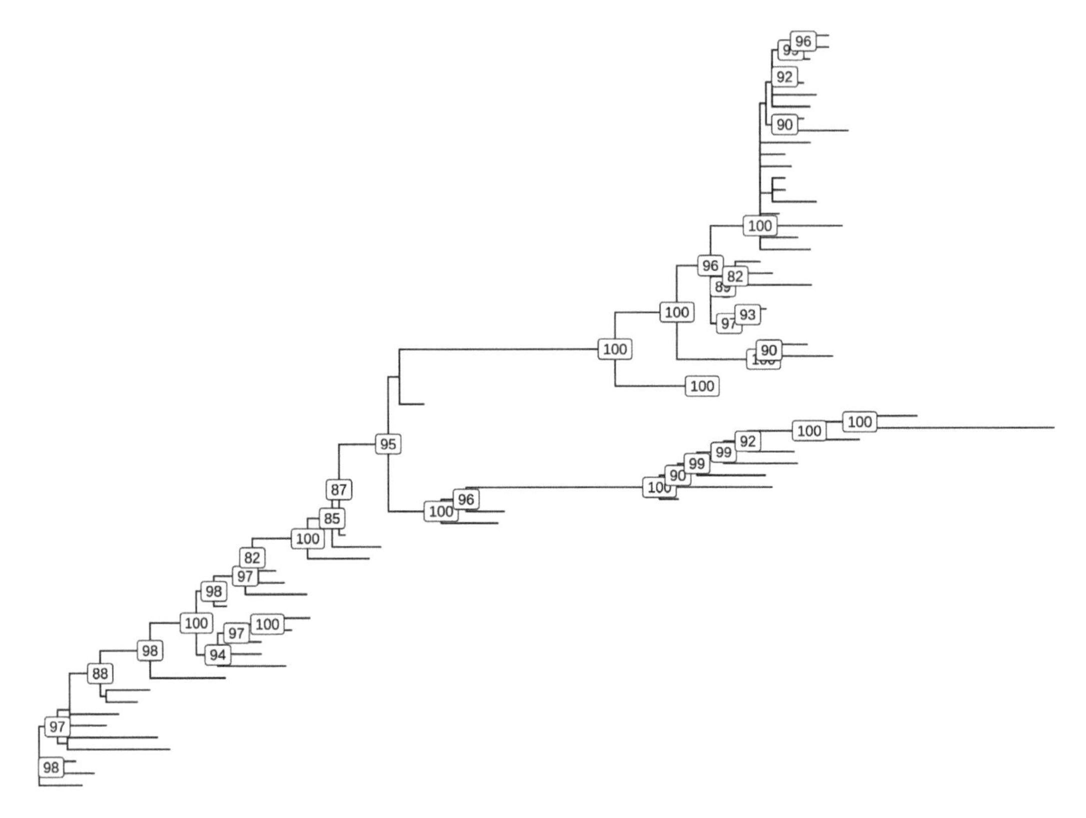

Figure A.5: **Bootstrap value stored in node label.**

As this is a very common issue, we implemented a `read.newick()` function in the **treeio** package to allow parsing internal node labels as supported values. As a result, it can be easier to display bootstrap values using the following code:

```
tr <- read.newick(nwk, node.label='support')
ggtree(tr) + geom_nodelab(geom='label', aes(label=support,
                          subset=support > 80))
```

A.5 Branch Setting

A.5.1 Plot the same tree as in `plot.phylo()`

By default, `ggtree()` ladderizes the input tree so that the tree will appear less cluttered. This is the reason why the tree visualized by `ggtree()` is different from the one using `plot.phylo()` which displays a non-ladderized tree. To disable the ladderize effect, users can pass the parameter `ladderize = FALSE` to the `ggtree()` function as demonstrated in Figure A.6.

```
library(ape)
library(ggtree)
set.seed(42)
x <- rtree(5)
plot(x)
ggtree(x, ladderize = FALSE) + geom_tiplab()
ggtree(x) + geom_tiplab()
```

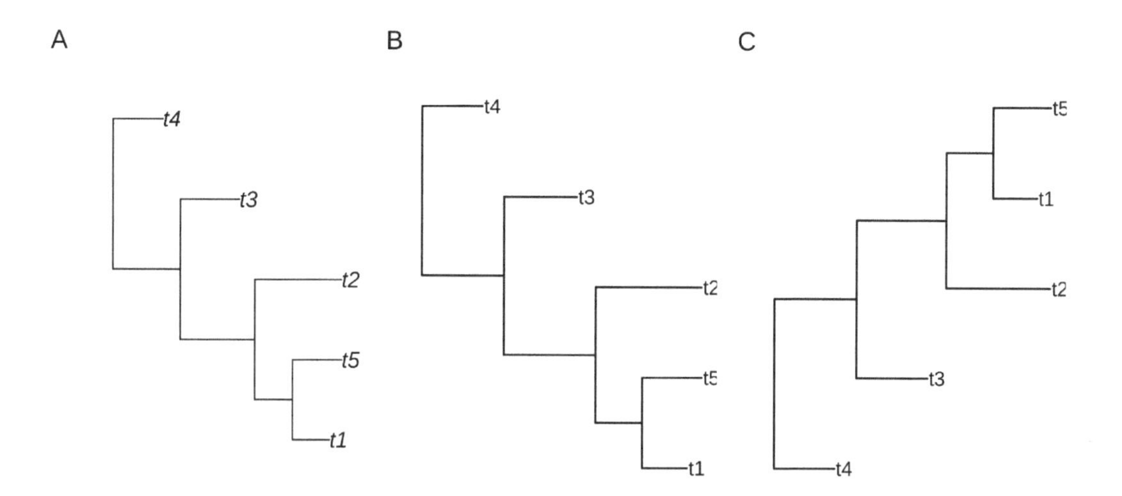

Figure A.6: **Ladderized and non-ladderized tree.** `plot.phylo()` displays non-ladderized tree (A), use `ladderize = FALSE` to display non-ladderized tree in `ggtree()` (B), `ggtree()` displays ladderized tree by default (C).

A.5.2 Specifying the order of the tips

The `rotateConstr()` function provided in the **ape** package rotates internal branches based on the specified order of the tips, and the order should be followed when plotting the tree (from bottom to top). As `ggtree()` by default ladderizes the input tree, users need to disable by passing `ladderize = FALSE`. Then the order of the tree will be displayed as expected (Figure A.7). Users can also extract tip order displayed by `ggtree()` using the `get_taxa_name()` function as demonstrated in session 12.6.

```
y <- ape::rotateConstr(x, c('t4', 't2', 't5', 't1', 't3'))
ggtree(y, ladderize = FALSE) + geom_tiplab()
```

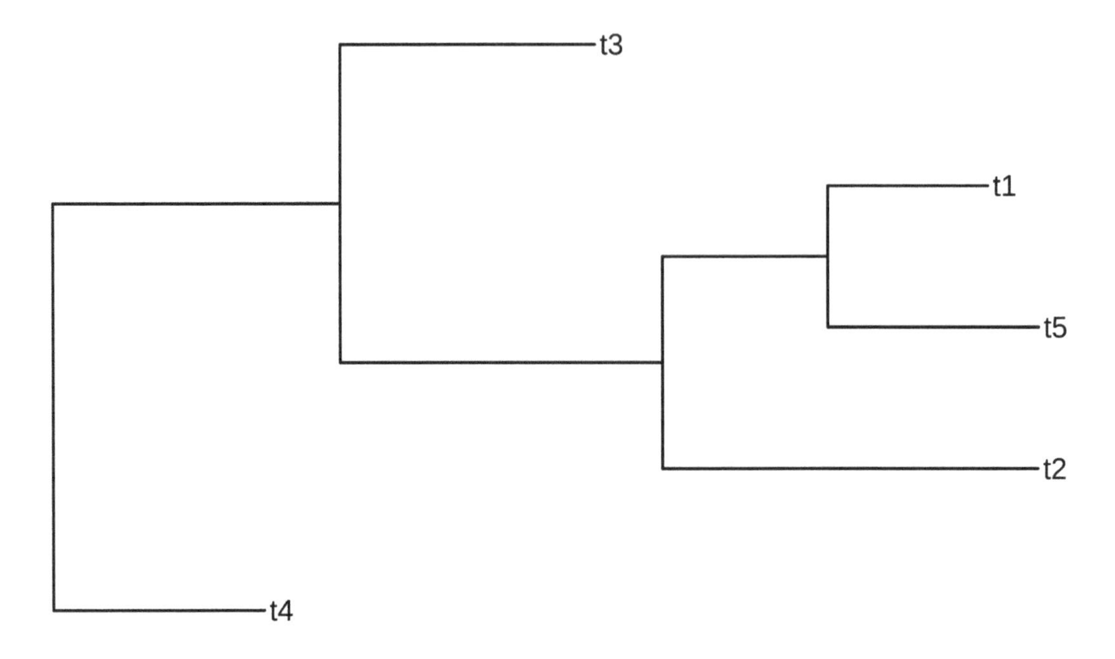

Figure A.7: **Specifying tree order.** The order of the input tree will be maintained in `ggtree()` when `ladderize = FALSE`.

A.5.3 Shrink outlier long branch

When outgroups are on a very long branch length (Figure A.8A), we would like to keep the outgroups in the tree but ignore their branch lengths (Figure A.8B)[5]. This can be easily done by modifying the coordinates of the outgroups (Figure A.8B). Another approach is to truncate the plot using the **ggbreak** package (Figure A.8C) (Xu, Chen, et al., 2021).

```
library(TDbook)
library(ggtree)

x <- tree_long_branch_example
m <- MRCA(x, 75, 76)
y <- groupClade(x, m)

## A
p <- p1 <- ggtree(y, aes(linetype = group)) +
  geom_tiplab(size = 2) +
  theme(legend.position = 'none')

## B
p$data[p$data$node %in% c(75, 76), "x"] <- mean(p$data$x)
```

[5]Example from: https://groups.google.com/d/msg/bioc-ggtree/T2ySvqv351g/mHsyljvBCwAJ

```
## C
library(ggbreak)
p2 <- p1 + scale_x_break(c(0.03, 0.09)) + hexpand(.05)

## align plot
plot_list(p1, p, p2, ncol=3, tag_levels="A")
```

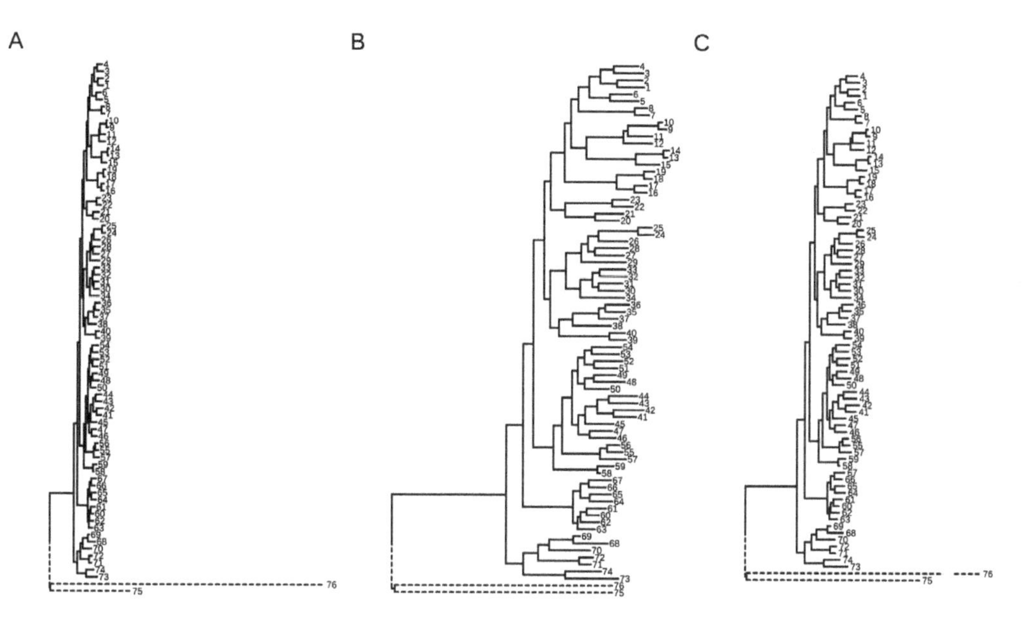

Figure A.8: **Shrink outlier long branch.** Original tree (A); reduced outgroup branch length (B); truncated tree plot (C).

A.5.4 Attach a new tip to a tree

Sometimes there are known branches that are not in the tree, but we would like to have them on the tree. Another common scenario is when we have a new sequence species and would like to update the reference tree with this species by inferring its evolutionary position.

Users can use `phytools::bind.tip()` (Revell, 2012) to attach a new tip to a tree. With **tidytree**, it is easy to add an annotation to differentiate newly introduced and original branches and to reflect the uncertainty of the added branch splits off, as demonstrated in Figure A.9.

```
library(phytools)
library(tidytree)
library(ggplot2)
library(ggtree)
```

```
set.seed(2019-11-18)
tr <- rtree(5)

tr2 <- bind.tip(tr, 'U', edge.length = 0.1, where = 7, position=0.15)
d <- as_tibble(tr2)
d$type <- "original"
d$type[d$label == 'U'] <- 'newly introduced'
d$sd <- NA
d$sd[parent(d, 'U')$node] <- 0.05

tr3 <- as.treedata(d)
ggtree(tr3, aes(linetype=type)) +  geom_tiplab() +
  geom_errorbarh(aes(xmin=x-sd, xmax=x+sd, y = y - 0.3),
                 linetype='dashed', height=0.1) +
  scale_linetype_manual(values = c("newly introduced" = "dashed",
                                   "original" = "solid")) +
  theme(legend.position=c(.8, .2))
```

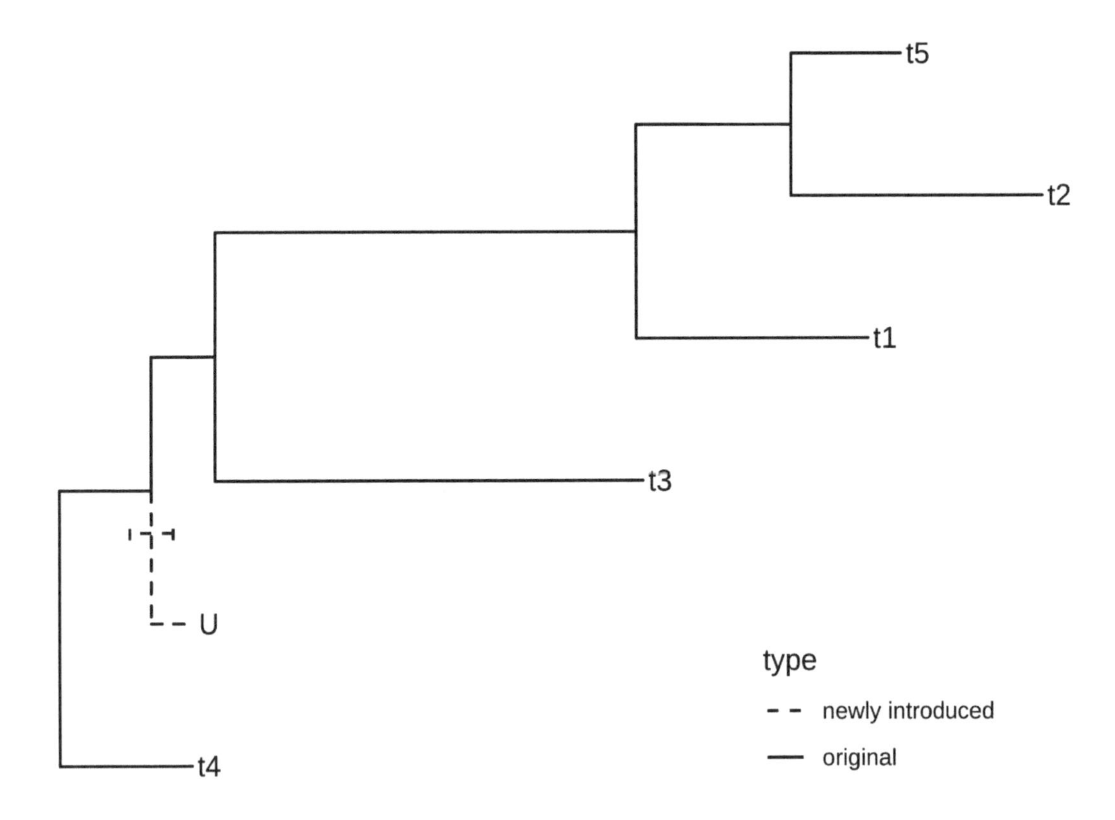

Figure A.9: **Attaching a new tip to a tree.** Different line types were employed to distinguish the newly introduced tip and an error bar was added to indicate the uncertainty of the added branch position.

A.5.5 Change colors or line types of arbitrarily selected branches

If you want to color or change line types of specific branches, you only need to prepare a data frame with variables of branch setting (e.g., selected and unselected). Applying the Method 1 described in (Yu et al., 2018) to map the data onto the tree will make it easy to set colors and line types (Figure A.10).

```
set.seed(123)
x <- rtree(10)
## binary choices of colors
d <- data.frame(node=1:Nnode2(x), colour = 'black')
d[c(2,3,14,15), 2] <- "red"

## multiple choices of line types
d2 <- data.frame(node=1:Nnode2(x), lty = 1)
d2[c(2,5,13, 14), 2] <- c(2, 3, 2,4)

p <- ggtree(x) + geom_label(aes(label=node))
p %<+% d %<+% d2 + aes(colour=I(colour), linetype=I(lty))
```

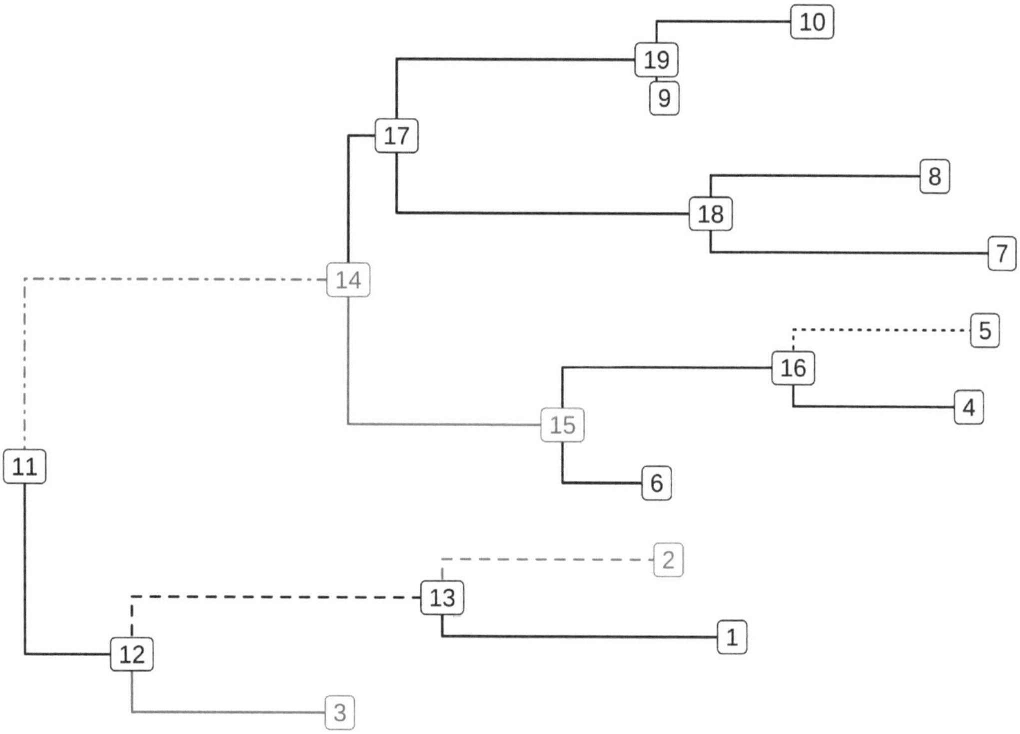

Figure A.10: **Change colors and line types of specific branches.**

Users can use the **gginnards** package to manipulate plot elements for more complicated scenarios.

A.5.6 Add an arbitrary point to a branch

If you want to add an arbitrary point to a branch[6], you can use `geom_nodepoint()`, `geom_tippoint()`, or `geom_point2()` (works for both external and internal nodes) to filter selected node (the endpoint of the branch) via the `subset` aesthetic mapping and specify horizontal position by `x = x - offset` aesthetic mapping, where the offset can be an absolute value (Figure A.11A) or in proportion to the branch length (Figure A.11B).

```
set.seed(2020-05-20)
x <- rtree(10)
p <- ggtree(x)

p1 <- p + geom_nodepoint(aes(subset = node == 13, x = x - .1),
                         size = 5, colour = 'firebrick', shape = 21)

p2 <- p + geom_nodepoint(aes(subset = node == 13,
                             x = x - branch.length * 0.2),
                         size = 3, colour = 'firebrick') +
       geom_nodepoint(aes(subset = node == 13, x = x - branch.length * 0.8),
                         size = 5, colour = 'steelblue')
plot_list(p1, p2, ncol=2, tag_levels="A")
```

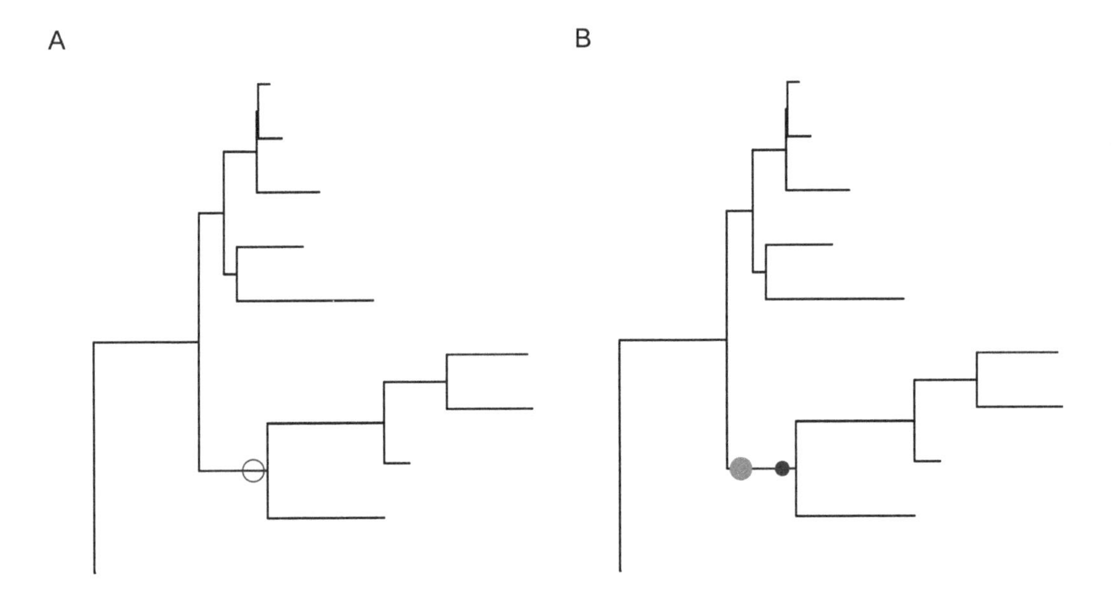

Figure A.11: **Add an arbitrary point on a branch.** The position of the symbolic point can be adjusted by an absolute value (A) or in proportion to the branch length (B).

[6]https://twitter.com/melanoidin/status/1262703932993871874

A.6 Different *X*-axis Labels for Different Facet Panels

This is not supported by **ggplot2** in general. However, we can just draw text labels for each panel and put the labels beyond the plot panels as demonstrated in Figure A.12.

```
library(ggtree)
library(ggplot2)
set.seed(2019-05-02)
x <- rtree(30)
p <- ggtree(x) + geom_tiplab()
d <- data.frame(label = x$tip.label,
                value = rnorm(30))
p2 <- p + geom_facet(panel = "Dot", data = d,
            geom = geom_point, mapping = aes(x = value))

p2 <- p2 + theme_bw() +
    xlim_tree(5) + xlim_expand(c(-5, 5), 'Dot')

# .panel is the internal variable used in `geom_facet` for faceting.
d <- data.frame(.panel = c('Tree', 'Dot'),
                lab = c("Distance", "Dot Units"),
                x=c(2.5,0), y=-2)

p2 + scale_y_continuous(limits=c(0, 31),
                        expand=c(0,0),
                        oob=function(x, ...) x) +
    geom_text(aes(label=lab), data=d) +
    coord_cartesian(clip='off')  +
    theme(plot.margin=margin(6, 6, 40, 6))
```

A.7 Plot Something behind the Phylogeny

The `ggtree()` function plots the tree structure, and normally we add layers on top of the tree.

```
set.seed(1982)
x <- rtree(5)
p <- ggtree(x) + geom_hilight(node=7, alpha=1)
```

If we want the layers behind the tree layer, we can reverse the order of all the layers.

```
p$layers <- rev(p$layers)
```

Another solution is to use `ggplot()` instead of `ggtree()` and `+ geom_tree()` to add the layer of tree structure at the correct position of the layer stack (Figure

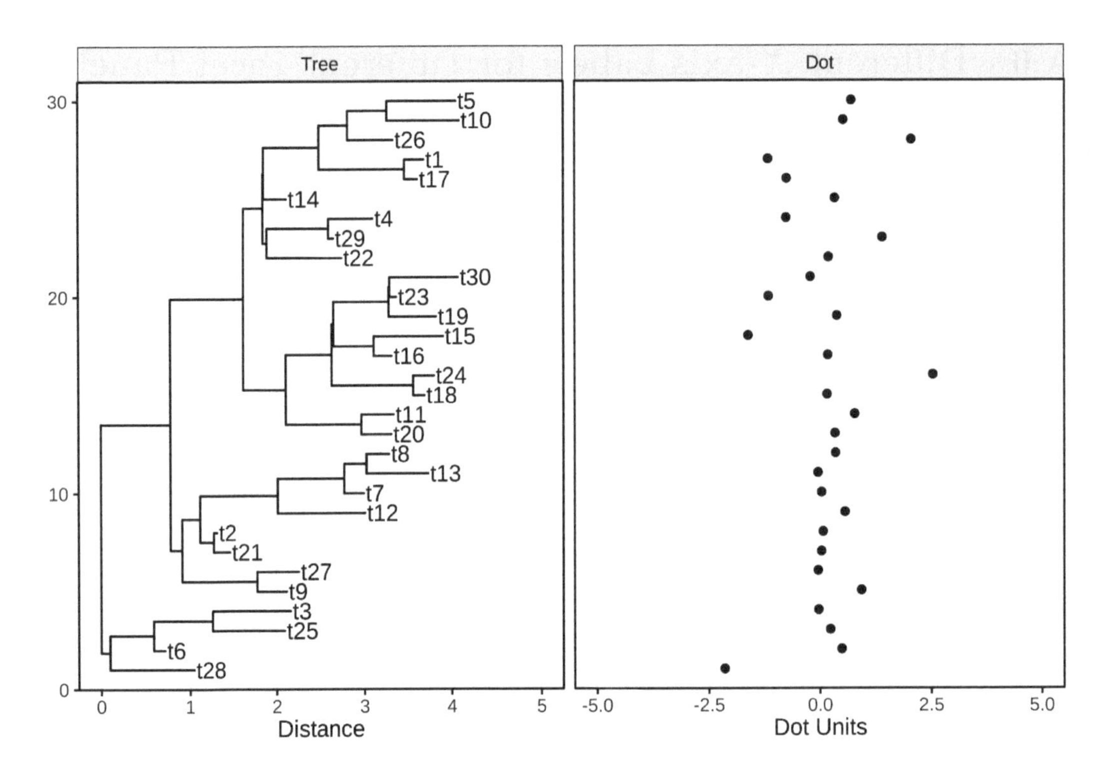

Figure A.12: ***X*-axis titles for different facet panels.**

A.13).

```
ggplot(x) + geom_hilight(node=7, alpha=1) + geom_tree() + theme_tree()
```

A.8 Enlarge Center Space in Circular/Fan Layout Tree

This question for enlarging center space in circular/fan layout tree was asked several times[7], and a published example can be found in (Barton et al., 2016). Increasing the percentage of center white space in a circular tree is useful to avoid overlapping tip labels and to increase the readability of the tree by moving all nodes and branches further out. This can be done simply by using `xlim()` or `hexpand()` to allocate more space (Figure A.14A), just like in Figure 4.3G, or assigning a long root branch that is similar to the "Root Length" parameter in **FigTree** (Figure A.14B).

```
set.seed(1982)
tree <- rtree(30)
plot_list(
  ggtree(tree, layout='circular') + xlim(-10, NA),
  ggtree(tree, layout='circular') + geom_rootedge(5),
```

[7]https://groups.google.com/d/msg/bioc-ggtree/gruC4FztU8I/mwavqWCXAQAJ, https://groups.google.com/d/msg/bioc-ggtree/UoGQekWHIvw/ZswUUZKSGwAJ and https://github.com/GuangchuangYu/ggtree/issues/95

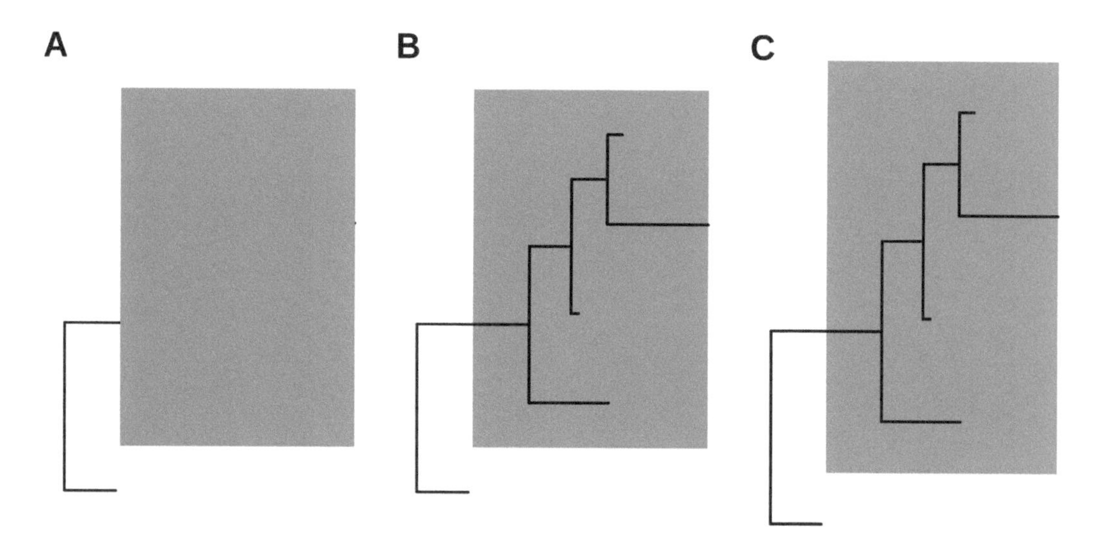

Figure A.13: **Add layers behind the tree structure.** A layer on top of the tree structure (A). Reverse layer order of A (B). Add layer behind the tree layer (C).

```
  tag_levels = "A", ncol=2
)
```

A.9 Use the Most Distant Tip from the Root as the Origin of the Timescale

The `revts()` will reverse the x-axis by setting the most recent tip to 0. We can use `scale_x_continuous(labels=abs)` to label x-axis using absolute values (Figure A.15).

```
tr <- rtree(10)
p <- ggtree(tr) + theme_tree2()
p2 <- revts(p) + scale_x_continuous(labels=abs)
plot_list(p, p2, ncol=2, tag_levels="A")
```

A.10 Remove Blank Margins for Circular Layout Tree

For plots in polar coordinates, such as a circular layout tree, it is very common that extra spaces will be generated.

If you are using `Rmarkdown`, you can set the following options for **knitr** to remove extra white space automatically.

```
library(knitr)
knit_hooks$set(crop = hook_pdfcrop)
```

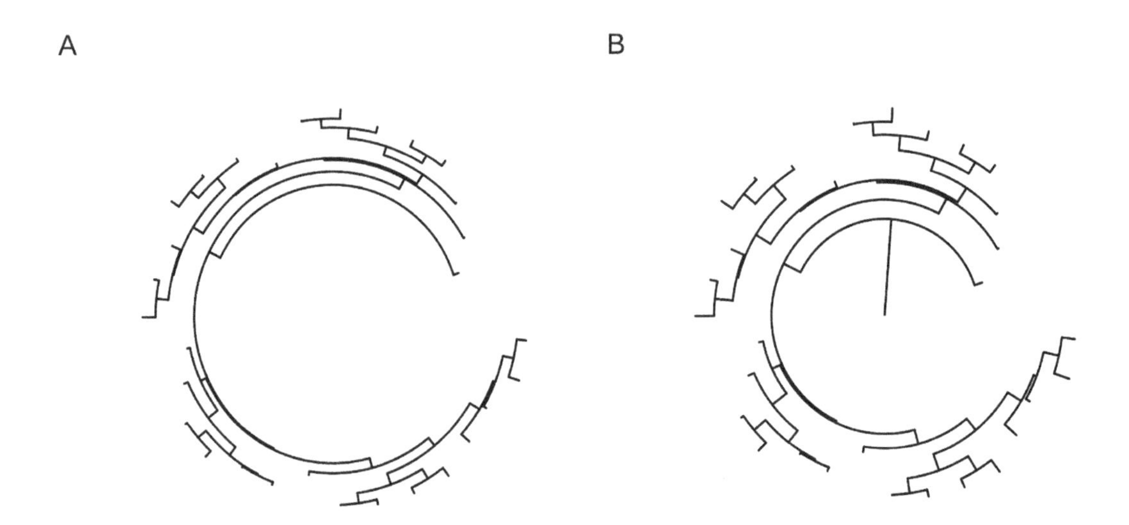

Figure A.14: **Enlarge center space in circular tree.** Allocate more space by `xlim` (A) or long root branch (B).

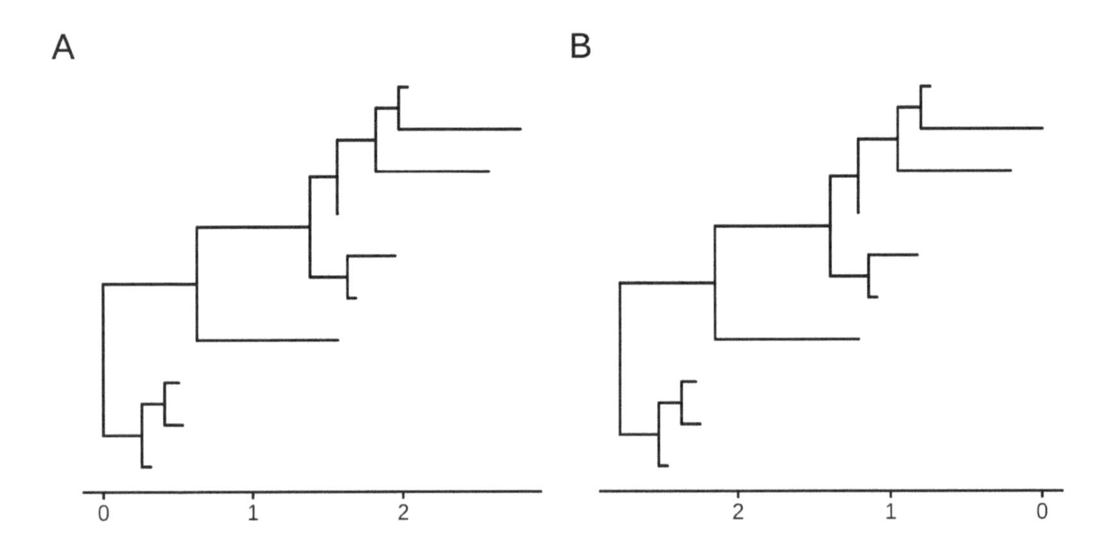

Figure A.15: **Origin of the time scale.** Forward: from the root to the tips (A). Backward: from the most distant tip to the root (B).

```
opts_chunk$set(crop = TRUE)
```

Otherwise, we can use command-line tools to remove extra white space:

```
## for pdf
pdfcrop x.pdf

## for png
convert -trim x.png x-crop.png
```

If you want to do it in R, you can use the **magick** package:

```
library(magick)

x <- image_read("x.png")
## x <- image_read_pdf("x.pdf") # for PDF

image_trim(x)
```

Here is an example (Figure A.16):

```
library(ggplot2)
library(ggtree)
library(patchwork)
library(magick)

set.seed(2021)
tr <- rtree(30)
p <- ggtree(tr, size=1, colour="purple", layout='circular')

f <- tempfile(fileext=".png")
ggsave(filename = f, plot = p, width=7, height=7)

x <- image_read(f, density=300)
y <- image_trim(x)

panel_border <- theme(panel.border=element_rect(colour='black',
                                               fill=NA, size=2))
xx <- image_ggplot(x) + panel_border
yy <- image_ggplot(y) + panel_border

plot_list(xx, yy, tag_levels = "A", ncol=2)
```

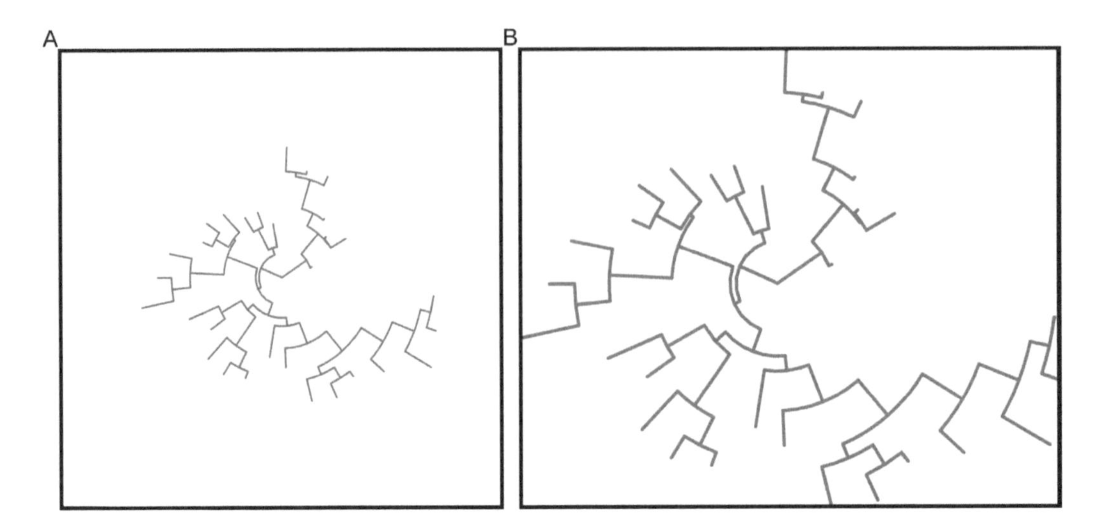

Figure A.16: **Trim extra white space for polar coordinates.** Original plot (A). Trimmed version (B).

A.11 Edit Tree Graphic Details

It can be hard to modify plot details for ordinary users using **ggplot2**/**ggtree**. We recommend using the **eoffice** package to export **ggtree** output to a Microsoft Office Document and edit the tree graphic in PowerPoint.

Appendix B

Related Tools

B.1 MicrobiotaProcess: Convert Taxonomy Table to a `treedata` Object

Taxonomy (genus, family, ...) data are widely used in microbiome or ecology. Hierarchical taxonomies are the tree-like structure that organizes items into subcategories and can be converted to a tree object (see also the phylog object). The **MicrobiotaProcess** supports converting a `taxonomyTable` object, defined in the **phyloseq** package, to a `treedata` object, and the taxonomic hierarchical relationship can be visualized using **ggtree** (Figure B.1). When there are taxonomy names that are confused and missing, the `as.treedata()` method for `taxonomyTable` objects will complete their upper-level taxonomic information automatically.

```
library(MicrobiotaProcess)
library(ggtree)

# The original kostic2012crc is a MPSE object
data(kostic2012crc)

taxa <- tax_table(kostic2012crc)
#The rownames (usually is OTUs or other features ) of the taxa will be
# served as the tip labels if include.rownames = TRUE
tree <- as.treedata(taxa, include.rownames=TRUE)
# Or extract the taxa tree (treedata) with mp_extract_tree, because the
# taxonomy information is stored as treedata in the MPSE class
# (kostic2012crc).
# tree <- kostic2012crc %>% mp_extract_tree()

ggtree(tree, layout="circular", size=0.2) +
    geom_tiplab(size=1)
```

Figure B.1: **Convert a `taxonomyTable` object to a `treedata` object.**

B.2 rtol: An R Interface to Open Tree API

The **rtol** (Michonneau et al., 2016) is an R package to interact with the Open Tree of Life data APIs. Users can use it to query phylogenetic trees and visualize the trees with **ggtree** to explore species relationships (Figure B.2).

```
## example from: https://github.com/ropensci/rotl
library(rotl)
apes <- c("Pongo", "Pan", "Gorilla", "Hoolock", "Homo")
(resolved_names <- tnrs_match_names(apes))

##   search_string unique_name approximate_match ott_id
## 1         pongo       Pongo             FALSE 417949
## 2           pan         Pan             FALSE 417957
```

```
## 3      gorilla     Gorilla                FALSE 417969
## 4      hoolock     Hoolock                FALSE 712902
## 5         homo        Homo                FALSE 770309
##   is_synonym          flags number_matches
## 1      FALSE                             2
## 2      FALSE sibling_higher              2
## 3      FALSE sibling_higher              1
## 4      FALSE                             1
## 5      FALSE sibling_higher              1
```

```
tr <- tol_induced_subtree(ott_ids = ott_id(resolved_names))
ggtree(tr) + geom_tiplab() + xlim(NA, 5)
```

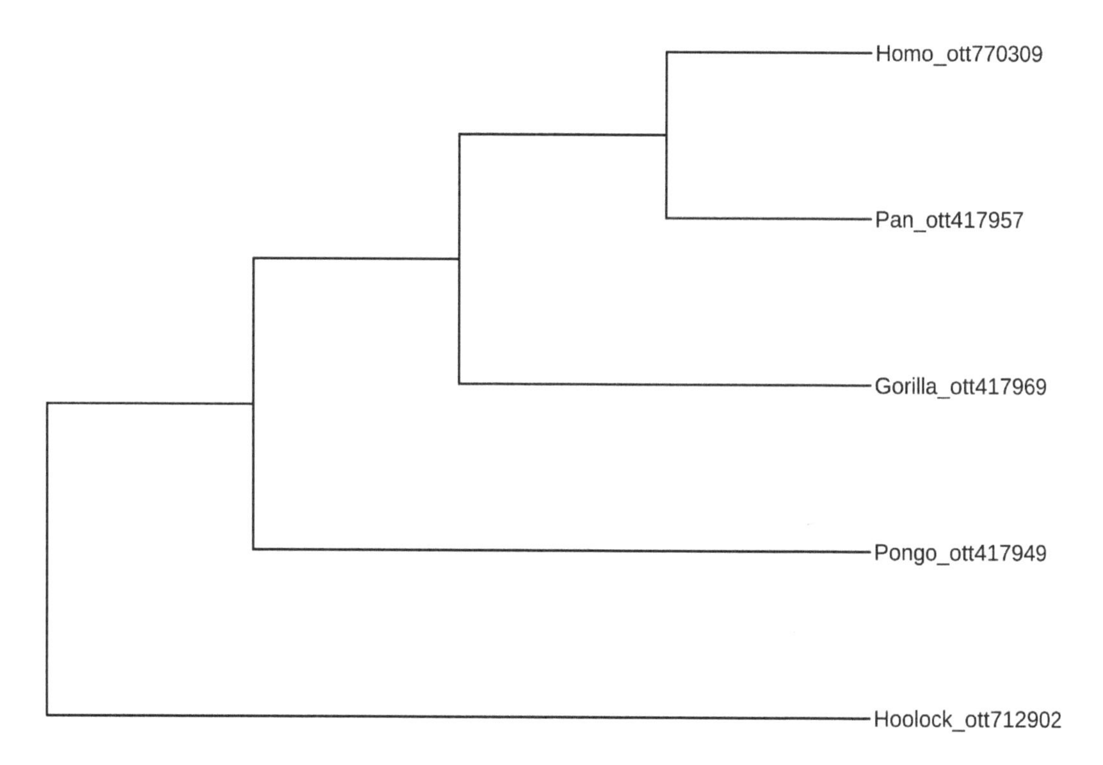

Figure B.2: **Get an induced subtree from the big Open Tree.**

B.3 Print ASCII-art Rooted Tree

```
library(data.tree)
tree <- rtree(10)
d <- as.data.frame(as.Node(tree))
names(d) <- NULL
print(d, row.names=FALSE)
```

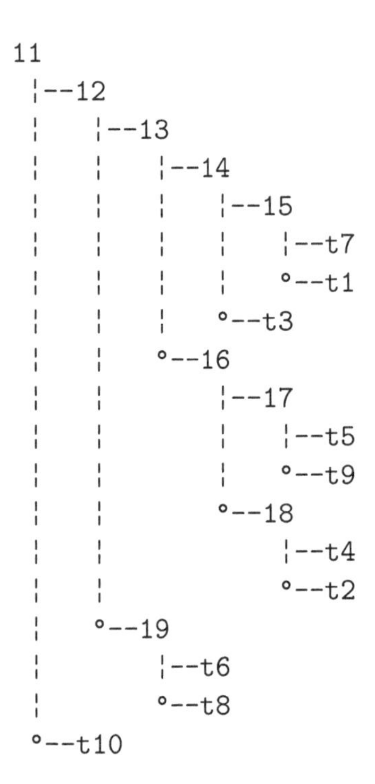

```
 11
  ¦--12
  ¦   ¦--13
  ¦   ¦   ¦--14
  ¦   ¦   ¦   ¦--15
  ¦   ¦   ¦   ¦   ¦--t7
  ¦   ¦   ¦   ¦   °--t1
  ¦   ¦   ¦   °--t3
  ¦   ¦   °--16
  ¦   ¦       ¦--17
  ¦   ¦       ¦   ¦--t5
  ¦   ¦       ¦   °--t9
  ¦   ¦       °--18
  ¦   ¦           ¦--t4
  ¦   ¦           °--t2
  ¦   °--19
  ¦       ¦--t6
  ¦       °--t8
  °--t10
```

It is neat to print ASCII-art of the phylogeny. Sometimes, we don't want to plot the tree, but just take a glance at the tree structure without leaving the focus from the R console. However, it is not a good idea to print the whole tree as ASCII text if the tree is large. Sometimes, we just want to look at a specific portion of the tree and its immediate relatives. In this scenario, we can use `treeio::tree_subset()` function (see session 2.4) to extract selected portion of a tree. Then we can print ASCII-art of the tree subset to explore the evolutionary relationship of the species of our interest in the R console.

The **ggtree** supports parsing tip labels as emoji to create phylomoji. With the **data.tree** and **emojifont** packages, we can also print phylomoji as ASCII text (Figure B.3).

```
library(data.tree)
library(emojifont)

tt <- '((snail,mushroom),(((sunflower,evergreen_tree),leaves),green_salad));'
tree <- read.tree(text = tt)
tree$tip.label <- emoji(tree$tip.label)
d <- as.data.frame(as.Node(tree))
names(d) <- NULL
print(d, row.names=FALSE)
```

Another way to print ASCII-art of phylogeny is to use the `ascii()` device defined in the **devout** package. Here is an example:

```
>
>
> library(data.tree)
> library(emojifont)
> tt <- '((snail,mushroom),(((sunflower,evergreen_tree),leaves),green_salad));'
> tree <- read.tree(text = tt)
> tree$tip.label <- emoji(tree$tip.label)
> d <- as.data.frame(as.Node(tree))
> names(d) <- NULL
> print(d, row.names=FALSE)

 7
  ¦--8
  ¦   ¦--🐌
  ¦   °--🍄
  °--9
      ¦--10
      ¦   ¦--11
      ¦   ¦   ¦--🌻
      ¦   ¦   °--🌲
      ¦   °--🍃
      °--🥗
> 
```

Figure B.3: **Print phylomoji as ASCII text.**

```
library(devout)
ascii(width=80)
ggtree(rtree(5))
invisible(dev.off())
```

```
                                                   ###################
                                                   #
                     ###############################
                     #                             #
                     #                             ###############
                     ##
                     ##
                     ##
  ###########################################################
  #                  #
  #                  #
  #                  #
  #                  ####################
  #
  #
  #
  ###########
```

B.4 Zoom in on the Selected Portion

In addition to using `viewClade()` function, users can use the **ggforce** package to zoom in on a selected clade (Figure B.4).

```
set.seed(2019-08-05)
x <- rtree(30)
nn <- tidytree::offspring(x, 43, self_include=TRUE)
ggtree(x) + ggforce::facet_zoom(xy = node %in% nn)
```

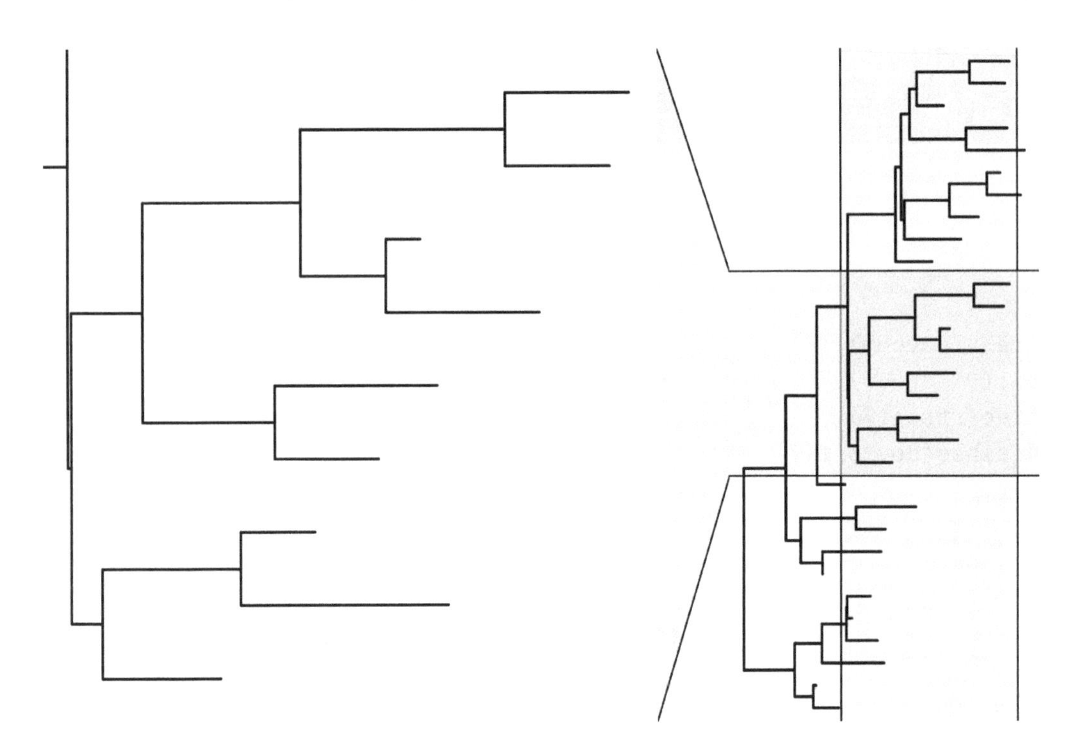

Figure B.4: **Zoom in on a selected clade.**

B.5 Tips for Using ggtree with ggimage

The **ggtree** supports annotating a tree with silhouette images via the **ggimage** package. The **ggimage** provides the grammar of graphic syntax to work with image files. It allows processing images on the fly via the `image_fun` parameter, which accepts a function to process `magick-image` objects (Figure B.5). The **magick** package provides several functions, and these functions can be combined to perform a particular task.

B.5.1 Example 1: Remove background of images

```
library(ggimage)

imgdir <- system.file("extdata/frogs", package = "TDbook")

set.seed(1982)
x <- rtree(5)
p <- ggtree(x) + theme_grey()
p1 <- p + geom_nodelab(image=paste0(imgdir, "/frog.jpg"),
                        geom="image", size=.12) +
      ggtitle("original image")
p2 <- p + geom_nodelab(image=paste0(imgdir, "/frog.jpg"),
            geom="image", size=.12,
            image_fun= function(.) magick::image_transparent(., "white")) +
      ggtitle("image with background removed")
plot_grid(p1, p2, ncol=2)
```

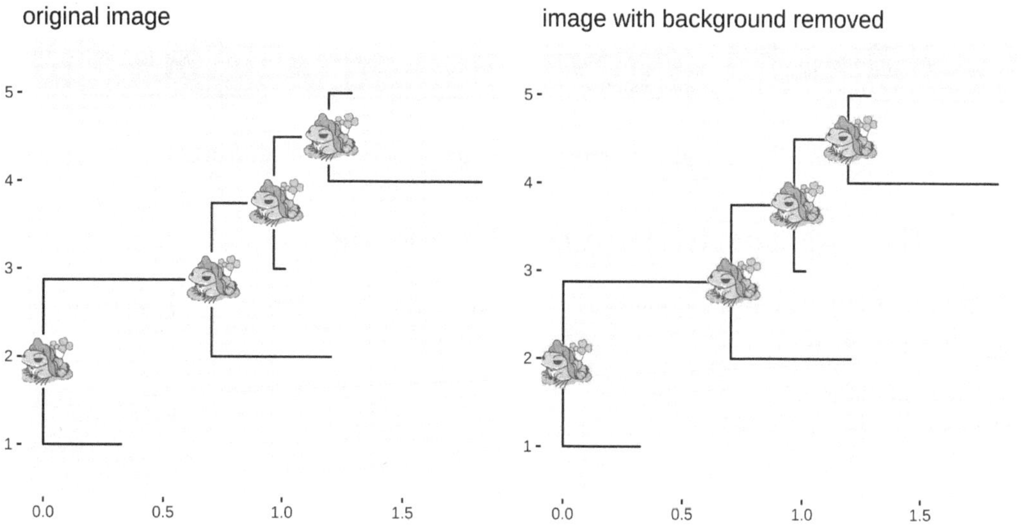

Figure B.5: **Remove image background.** Plotting silhouette images on a phylogenetic tree with background not removed (A) and removed (B).

B.5.2 Example 2: Plot tree on a background image

The `geom_bgimage()` adds a layer of the image and puts the layer to the bottom of the layer stack. It is a normal layer and doesn't change the structure of the output `ggtree` object. Users can add annotation layers without the background image layer (Figure B.6).

```
ggtree(rtree(20), size=1.5, color="white") +
  geom_bgimage('img/blackboard.jpg') +
  geom_tiplab(color="white", size=5, family='xkcd')
```

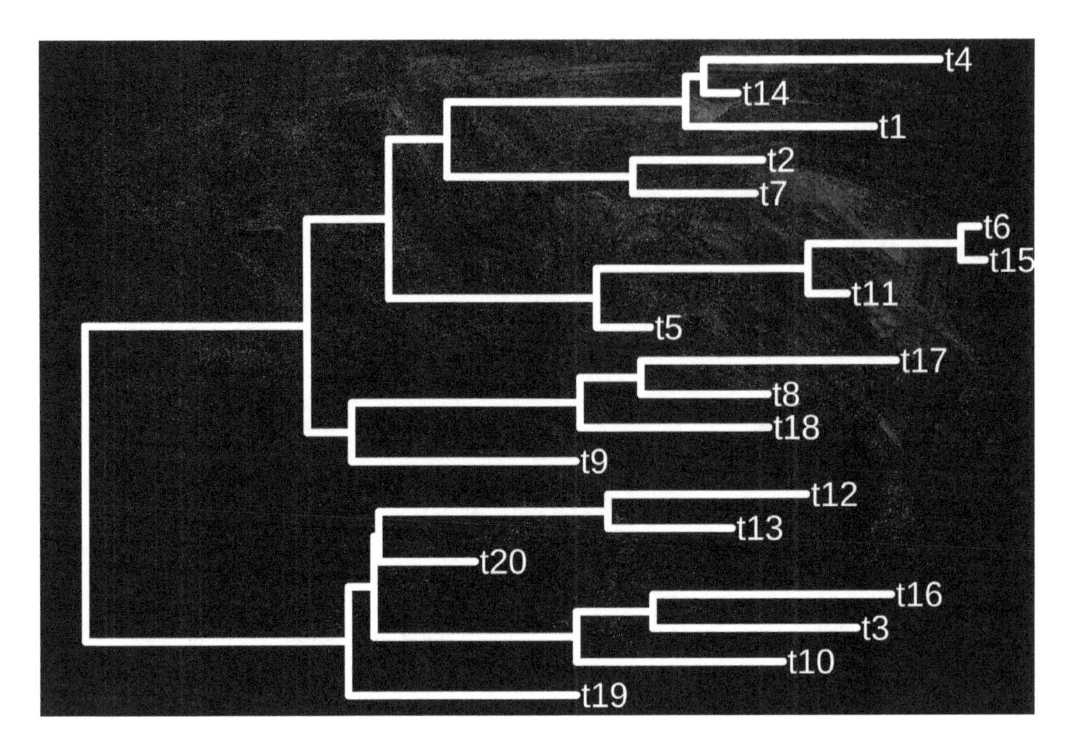

Figure B.6: **Use an image file as a tree background.**

B.6 Run ggtree in Jupyter Notebook

If you have Jupyter notebook installed on your system, you can install IRkernel with the following command in R:

```
install.packages("IRkernel")
IRkernel::installspec()
```

Then you can use **ggtree** and other R packages in the Jupyter notebook (Figure B.7). Here is a screenshot of recreating Figure 8.5 in the Jupyter notebook.

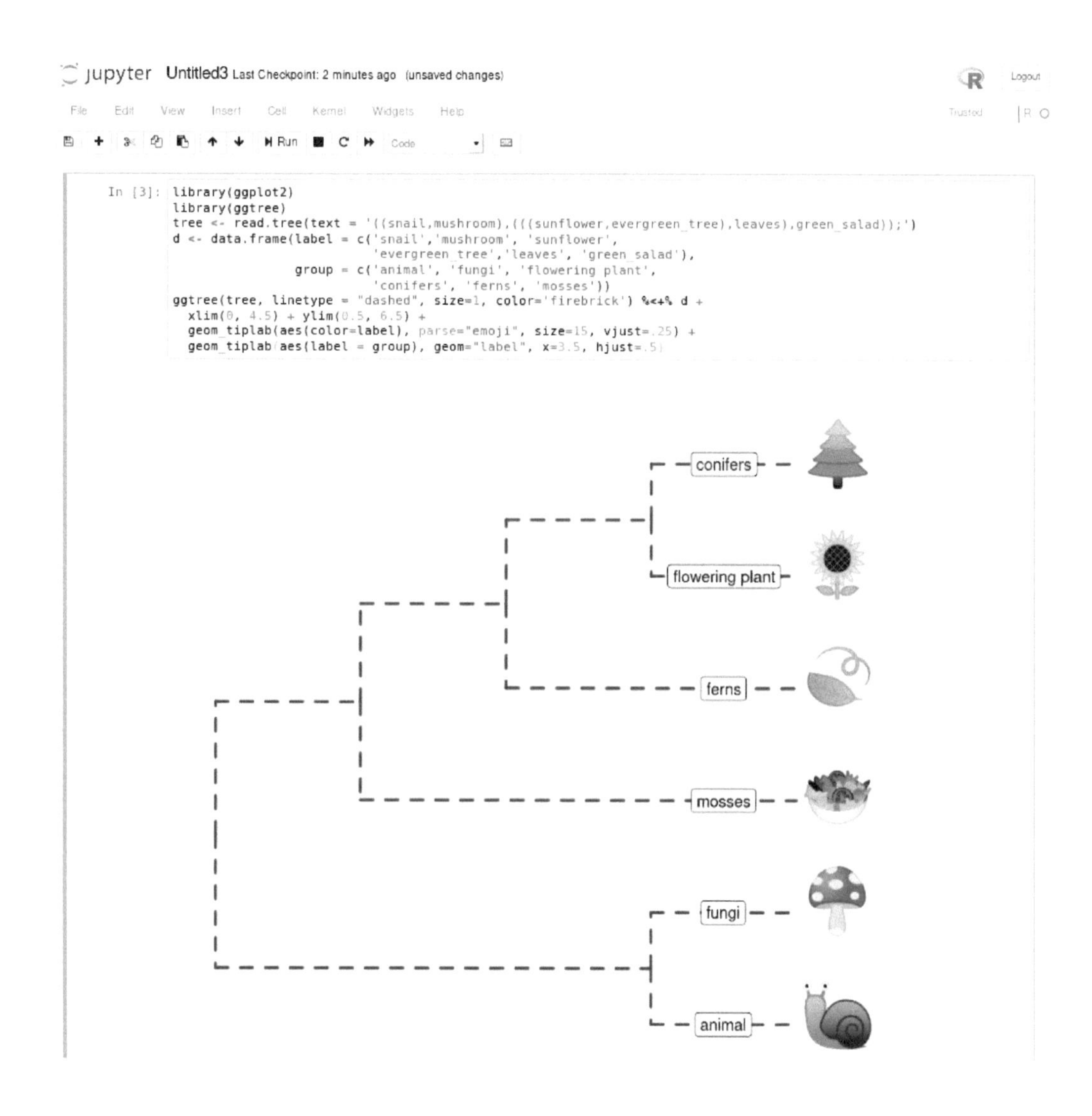

Figure B.7: **ggtree in Jupyter notebook.** Running ggtree in Jupyter notebook via R kernel.

Figures and Tables

Table 1: Geometric layers that supported by, 'geom_facet()'

Package	Geom Layer	Description
ggalt	geom_dumbbell	creates dumbbell charts
ggbio	geom_alignment	shows interval data as alignment
ggfittext	geom_fit_text	shrinks, grows, or wraps text to fit inside a defined rectangular area
gggenes	geom_gene_arrow	draws genes as arrows
ggimage	geom_image	visualizes image files
	geom_phylopic	queries image files from the PhyloPic database and visualizes them
ggplot2	geom_hline	adds horizontal lines
	geom_jitter	adds a small amount of random variation to the location of each point
	geom_label	draws a rectangle behind the text
	geom_point	creates scatterplots
	geom_raster	a high-performance special case for all the tiles that are the same size
	geom_rect	draws rectangle by using the locations of the four corners
	geom_segment	draws a straight line between points
	geom_spoke	a polar parameterization of 'geom_segment()'
	geom_text	adds text to the plot
	geom_tile	draws rectangle by using the center of the tile and its size
	geom_vline	adds vertical lines
ggrepel	geom_text_repel	adds text to the plot. The text labels repel away from each other and away from the data points
	geom_label_repel	draws a rectangle underneath the text. The text labels repel away from each other and away from the data points
ggridges	geom_density_ridges	arranges multiple density plots in a staggered fashion
	geom_density_ridges_gradient	works just like 'geom_density_ridges' except that the 'fill' aesthetic can vary along the *x*-axis
	geom_ridgeline	plots the sum of the 'y' and 'height' aesthetics vs. 'x', filling the area between 'y' and 'y + height' with a color
	geom_ridgeline_gradient	works just like 'geom_ridgeline' except that the 'fill' aesthetic can vary along the *x*-axis
ggstance	geom_barh	horizontal version of 'geom_bar()'
	geom_boxploth	horizontal version of 'geom_boxplot()'
	geom_crossbarh	horizontal version of 'geom_crossbar()'
	geom_errorbarh	horizontal version of 'geom_errorbarh()'
	geom_histogramh	horizontal version of 'geom_histogram()'
	geom_linerangeh	horizontal version of 'geom_linerange()'
	geom_pointrangeh	horizontal version of 'geom_pointrange()'
	geom_violinh	horizontal version of 'geom_violin()'
ggtree	geom_motif	draws aligned motifs

Table 2: Tree-like objects supported by ggtree

Package	Object	Description
ape	phylo	basic phylogenetic tree structure
	multiPhylo	list of phylo objects
ade4	phylog	tree structure for ecological data
phylobase	phylo4	S4 version of phylo object
	phylo4d	extend phylo4 with trait data
phyloseq	phyloseq	phylogenetic tree with microbiome data
tidytree	tbl_tree	phylogenetic tree as a tidy data frame
treeio	treedata	phylogenetic tree with heterogeneous associated data
	jplace	treedata object with placement information
stats	hclust	hierarchical cluster result
	dendrogram	hierarchical clustering or classification/regression tree
cluster	agnes	agglomerative hierarchical clustering
	diana	divisive hierarchical clustering
	twins	agglomerative or divisive (polythetic) hierarchical clustering
pvclust	pvclust	hierarchical clustering with p-values calculated by multiscale bootstrap resampling
igraph	igraph	network (currently only tree graph supported)

Publications of the ggtree Package Suite

1. S Xu, Z Dai, P Guo, X Fu, S Liu, L Zhou, W Tang, T Feng, M Chen, L Zhan, T Wu, E Hu, Y Jiang*, X Bo*, **G Yu***. ggtreeExtra: Compact visualization of richly annotated phylogenetic data. ***Molecular Biology and Evolution***. 2021, 38(9):4039-4042. doi: 10.1093/molbev/msab166
2. **G Yu**. Using ggtree to visualize data on tree-like structures. ***Current Protocols in Bioinformatics***, 2020, 69:e96. doi: 10.1002/cpbi.96
3. LG Wang, TTY Lam, S Xu, Z Dai, L Zhou, T Feng, P Guo, CW Dunn, BR Jones, T Bradley, H Zhu, Y Guan, Y Jiang, **G Yu***. treeio: an R package for phylogenetic tree input and output with richly annotated and associated data. ***Molecular Biology and Evolution***. 2020, 37(2):599-603. doi: 10.1093/molbev/msz240
4. **G Yu***, TTY Lam, H Zhu, Y Guan*. Two methods for mapping and visualizing associated data on phylogeny using ggtree. ***Molecular Biology and Evolution***. 2018, 35(2):3041-3043. doi: 10.1093/molbev/msy194
5. **G Yu**, DK Smith, H Zhu, Y Guan, TTY Lam*. ggtree: an R package for visualization and annotation of phylogenetic trees with their covariates and other associated data. ***Methods in Ecology and Evolution***. 2017, 8(1):28-36. doi: 10.1111/2041-210X.12628

Note: * Co-corresponding authors

References

Amer, A., Galvin, S., Healy, C. M., & Moran, G. P. (2017). The microbiome of potentially malignant oral leukoplakia exhibits enrichment for fusobacterium, leptotrichia, campylobacter, and rothia species. *Frontiers in Microbiology*, *8*, 2391. https://doi.org/10.3389/fmicb.2017.02391

Arenas, M. (2015). Trends in substitution models of molecular evolution. *Frontiers in Genetics*, *6*. https://doi.org/10.3389/fgene.2015.00319

Asnicar, F., Weingart, G., Tickle, T. L., Huttenhower, C., & Segata, N. (2015). Compact graphical representation of phylogenetic data and metadata with GraPhlAn. *PeerJ*, *3*, e1029. https://doi.org/10.7717/peerj.1029

Barton, K., Hiener, B., Winckelmann, A., Rasmussen, T. A., Shao, W., Byth, K., Lanfear, R., Solomon, A., McMahon, J., Harrington, S., Buzon, M., Lichterfeld, M., Denton, P. W., Olesen, R., Østergaard, L., Tolstrup, M., Lewin, S. R., Søgaard, O. S., & Palmer, S. (2016). Broad activation of latent HIV-1 in vivo. *Nature Communications*, *7*, 12731. https://doi.org/10.1038/ncomms12731

Berger, S. A., Krompass, D., & Stamatakis, A. (2011). Performance, Accuracy, and Web Server for Evolutionary Placement of Short Sequence Reads under Maximum Likelihood. *Systematic Biology*, 291–302. https://doi.org/10.1093/sysbio/syr010

Bosi, E., Monk, J. M., Aziz, R. K., Fondi, M., Nizet, V., & Palsson, B. Ø. (2016). Comparative genome-scale modelling of Staphylococcus aureus strains identifies strain-specific metabolic capabilities linked to pathogenicity. *Proceedings of the National Academy of Sciences of the United States of America*, *113*(26), E3801–E3809. https://doi.org/10.1073/pnas.1523199113

Bouckaert, R., Heled, J., Kühnert, D., Vaughan, T., Wu, C.-H., Xie, D., Suchard, M. A., Rambaut, A., & Drummond, A. J. (2014). BEAST 2: A Software Platform for Bayesian Evolutionary Analysis. *PLoS Comput Biol*, *10*(4), e1003537. https://doi.org/10.1371/journal.pcbi.1003537

Boussau, B., Szöllősi, G. J., Duret, L., Gouy, M., Tannier, E., & Daubin, V. (2013). Genome-scale coestimation of species and gene trees. *Genome Research*, *23*(2), 323–330. https://doi.org/10.1101/gr.141978.112

Callahan, B. J., McMurdie, P. J., Rosen, M. J., Han, A. W., Johnson, A. J. A., & Holmes, S. P. (2016). DADA2: High-resolution sample inference from Illumina amplicon data. *Nature Methods*, *13*(7), 581–583. https://doi.org/10.1038/nmeth.3869

Chen, Z., Ho, W. C. S., Boon, S. S., Law, P. T. Y., Chan, M. C. W., DeSalle, R., Burk, R. D., & Chan, P. K. S. (2017). Ancient evolution and dispersion of human papillomavirus 58 variants. *Journal of Virology*, *91*(21), e01285–17. https://doi.org/10.1128/JVI.01285-17

Chevenet, F., Brun, C., Bañuls, A.-L., Jacq, B., & Christen, R. (2006). TreeDyn: Towards dynamic graphics and annotations for analyses of trees. *BMC Bioinformatics*, *7*, 439. https://doi.org/10.1186/1471-2105-7-439

Chow, N. A., Muñoz, J. F., Gade, L., Berkow, E. L., Li, X., Welsh, R. M., Forsberg, K., Lockhart, S. R., Adam, R., Alanio, A., Alastruey-Izquierdo, A., Althawadi, S., Araúz, A. B., Ben-Ami, R., Bharat, A., Calvo, B., Desnos-Ollivier, M., Escandón, P., Gardam, D., … Cuomo, C. A. (2020). Tracing the evolutionary history and global expansion of candida auris using population genomic analyses. *mBio*, *11*(2). https://doi.org/10.1128/mBio.03364-19

Czech, L., Huerta-Cepas, J., & Stamatakis, A. (2017). A Critical Review on the Use of Support Values in Tree Viewers and Bioinformatics Toolkits. *Molecular Biology and Evolution*, *34*(6), 1535–1542. https://doi.org/10.1093/molbev/msx055

Escudero, M., & Wendel, J. F. (2020). The grand sweep of chromosomal evolution in angiosperms. *New Phytologist*, *228*(3), 805–808. https://doi.org/10.1111/nph.16802

Felsenstein, J. (1978). Cases in which Parsimony or Compatibility Methods will be Positively Misleading. *Systematic Biology*, *27*(4), 401–410. https://doi.org/10.1093/sysbio/27.4.401

Felsenstein, J. (1981). Evolutionary trees from DNA sequences: A maximum likelihood approach. *Journal of Molecular Evolution*, *17*(6), 368–376.

Felsenstein, J. (1989). PHYLIP - Phylogeny Inference Package (Version 3.2). *Cladistics*, *5*, 164–166.

Fitch, W. M. (1971). Toward Defining the Course of Evolution: Minimum Change for a Specific Tree Topology. *Systematic Zoology*, *20*(4), 406–416. https://doi.org/10.2307/2412116

Gentleman, R. C., Carey, V. J., Bates, D. M., Bolstad, B., Dettling, M., Dudoit, S., Ellis, B., Gautier, L., Ge, Y., Gentry, J., Hornik, K., Hothorn, T., Huber, W., Iacus, S., Irizarry, R., Leisch, F., Li, C., Maechler, M., Rossini, A. J., … Zhang, J. (2004). Bioconductor: Open software development for computational biology and bioinformatics. *Genome Biology*, *5*(10), R80. https://doi.org/10.1186/gb-2004-5-10-r80

Goldman, N., & Yang, Z. (1994). A codon-based model of nucleotide substitution for protein-coding DNA sequences. *Molecular Biology and Evolution*, *11*(5), 725–736.

Grubaugh, N. D., Ladner, J. T., Kraemer, M. U. G., Dudas, G., Tan, A. L., Gangavarapu, K., Wiley, M. R., White, S., Thézé, J., Magnani, D. M., Prieto, K., Reyes, D., Bingham, A. M., Paul, L. M., Robles-Sikisaka, R., Oliveira, G., Pronty, D., Barcellona, C. M., Metsky, H. C., … Andersen, K. G. (2017). Genomic epidemiology reveals multiple introductions of zika virus into the united states. *Nature*, *546*(7658), 401–405. https://doi.org/10.1038/nature22400

Gupta, A., & Sharma, V. K. (2015). Using the taxon-specific genes for the taxonomic classification of bacterial genomes. *BMC Genomics*, *16*(1). https://doi.org/10.1186/s12864-015-1542-0

He, Y.-Q., Chen, L., Xu, W.-B., Yang, H., Wang, H.-Z., Zong, W.-P., Xian, H.-X., Chen, H.-L., Yao, X.-J., Hu, Z.-L., Luo, M., Zhang, H.-L., Ma, H.-W., Cheng, J.-Q., Feng, Q.-J., & Zhao, D.-J. (2013). Emergence, Circulation, and Spatiotemporal Phylogenetic Analysis of Coxsackievirus A6- and Coxsackievirus A10-Associated Hand, Foot, and Mouth Disease Infections from 2008 to 2012 in Shenzhen, China. *Journal of Clinical Microbiology*, *51*(11), 3560–3566. https://doi.org/10.1128/JCM.01231-13

He, Z., Gharaibeh, R. Z., Newsome, R. C., Pope, J. L., Dougherty, M. W., Tomkovich, S., Pons, B., Mirey, G., Vignard, J., Hendrixson, D. R., & Jobin, C. (2019). Campylobacter jejuni promotes colorectal tumorigenesis through the action of cytolethal distending toxin. *Gut*, *68*(2), 289–300. https://doi.org/10.1136/gutjnl-2018-317200

He, Z., Zhang, H., Gao, S., Lercher, M. J., Chen, W.-H., & Hu, S. (2016). Evolview v2: An online visualization and management tool for customized and annotated phylogenetic trees. *Nucleic Acids Research*, *44*(W1), W236–241. https://doi.org/10.1093/nar/gkw370

Höhna, S., Heath, T. A., Boussau, B., Landis, M. J., Ronquist, F., & Huelsenbeck, J. P. (2014). Probabilistic graphical model representation in phylogenetics. *Systematic Biology*, *63*(5), 753–771. https://doi.org/10.1093/sysbio/syu039

Höhna, S., Landis, M. J., Heath, T. A., Boussau, B., Lartillot, N., Moore, B. R., Huelsenbeck, J. P., & Ronquist, F. (2016). RevBayes: Bayesian Phylogenetic Inference Using Graphical Models and an Interactive Model-Specification Language. *Systematic Biology*, *65*(4), 726–736. https://doi.org/10.1093/sysbio/syw021

Huelsenbeck, J. P., & Ronquist, F. (2001). MRBAYES: Bayesian inference of phylogenetic trees. *Bioinformatics (Oxford, England)*, *17*(8), 754–755.

Huson, D. H., & Scornavacca, C. (2012). Dendroscope 3: An interactive tool for rooted phylogenetic trees and networks. *Systematic Biology*, *61*(6), 1061–1067. https://doi.org/10.1093/sysbio/sys062

Jombart, T., Aanensen, D. M., Baguelin, M., Birrell, P., Cauchemez, S., Camacho, A., Colijn, C., Collins, C., Cori, A., Didelot, X., Fraser, C., Frost, S., Hens, N., Hugues, J., Höhle, M., Opatowski, L., Rambaut, A., Ratmann, O., Soubeyrand, S., … Ferguson, N. (2014). OutbreakTools: A new platform for disease outbreak analysis using the R software. *Epidemics*, *7*, 28–34. https://doi.org/10.1016/j.epidem.2014.04.003

Kostic, A. D., Gevers, D., Pedamallu, C. S., Michaud, M., Duke, F., Earl, A. M., Ojesina, A. I., Jung, J., Bass, A. J., Tabernero, J.others. (2012). Genomic analysis identifies association of fusobacterium with colorectal carcinoma. *Genome Research*, *22*(2), 292–298. https://doi.org/10.1101/gr.126573.111

Kuczynski, J., Stombaugh, J., Walters, W. A., González, A., Caporaso, J. G., & Knight, R. (2011). Using QIIME to analyze 16S rRNA gene sequences from Microbial Communities. *Current Protocols in Bioinformatics / Editoral Board,*

Andreas D. Baxevanis ... [Et Al.], CHAPTER, Unit10.7. https://doi.org/10.1002/0471250953.bi1007s36

Kumar, S., Stecher, G., & Tamura, K. (2016). MEGA7: Molecular evolutionary genetics analysis version 7.0 for bigger datasets. *Molecular Biology and Evolution*, *33*(7), 1870–1874. https://doi.org/10.1093/molbev/msw054

Kunin, V., & Hugenholtz, P. (2010). PyroTagger : A fast , accurate pipeline for analysis of rRNA amplicon pyrosequence data. *The Open Journal*, 1–8. http://www.theopenjournal.org/toj_articles/1

Lam, T. T.-Y., Hon, C.-C., Lemey, P., Pybus, O. G., Shi, M., Tun, H. M., Li, J., Jiang, J., Holmes, E. C., & Leung, F. C.-C. (2012). Phylodynamics of H5N1 avian influenza virus in Indonesia. *Molecular Ecology*, *21*(12), 3062–3077. https://doi.org/10.1111/j.1365-294X.2012.05577.x

Lam, T. T.-Y., Zhou, B., Wang, J., Chai, Y., Shen, Y., Chen, X., Ma, C., Hong, W., Chen, Y., Zhang, Y., Duan, L., Chen, P., Jiang, J., Zhang, Y., Li, L., Poon, L. L. M., Webby, R. J., Smith, D. K., Leung, G. M., ... Zhu, H. (2015). Dissemination, divergence and establishment of H7N9 influenza viruses in China. *Nature*, *522*(7554), 102–105. https://doi.org/10.1038/nature14348

Larsen, F. T., Bed'Hom, B., Guldbrandtsen, B., & Dalgaard, T. S. (2019). Identification and tissue-expression profiling of novel chicken c-type lectin-like domain containing proteins as potential targets for carbohydrate-based vaccine strategies. *Molecular Immunology*, *114*, 216–225. https://doi.org/10.1016/j.molimm.2019.07.022

Lemmon, A. R., & Moriarty, E. C. (2004). The importance of proper model assumption in bayesian phylogenetics. *Systematic Biology*, *53*(2), 265–277. https://doi.org/10.1080/10635150490423520

Letunic, I., & Bork, P. (2007). Interactive Tree Of Life (iTOL): An online tool for phylogenetic tree display and annotation. *Bioinformatics*, *23*(1), 127–128. https://doi.org/10.1093/bioinformatics/btl529

Liang, H., Lam, T. T.-Y., Fan, X., Chen, X., Zeng, Y., Zhou, J., Duan, L., Tse, M., Chan, C.-H., Li, L., Leung, T.-Y., Yip, C.-H., Cheung, C.-L., Zhou, B., Smith, D. K., Poon, L. L.-M., Peiris, M., Guan, Y., & Zhu, H. (2014). Expansion of genotypic diversity and establishment of 2009 H1N1 pandemic-origin internal genes in pigs in China. *Journal of Virology*, JVI.01327–14. https://doi.org/10.1128/JVI.01327-14

Lott, S. C., Voß, B., Hess, W. R., & Steglich, C. (2015). CoVennTree: A new method for the comparative analysis of large datasets. *Frontiers in Genetics*, *6*, 43. https://doi.org/10.3389/fgene.2015.00043

Maddison, D. R., Swofford, D. L., Maddison, W. P., & Cannatella, D. (1997). Nexus: An Extensible File Format for Systematic Information. *Systematic Biology*, *46*(4), 590–621. https://doi.org/10.1093/sysbio/46.4.590

Matsen, F. A., Hoffman, N. G., Gallagher, A., & Stamatakis, A. (2012). A Format for Phylogenetic Placements. *PLOS ONE*, *7*(2), e31009. https://doi.org/10.1371/journal.pone.0031009

Matsen, F. A., Kodner, R. B., & Armbrust, E. V. (2010). Pplacer: Linear time maximum-likelihood and bayesian phylogenetic placement of sequences onto a fixed reference tree. *BMC Bioinformatics*, *11*(1), 538. https://doi.org/10.1186/1471-2105-11-538

McMurdie, P. J., & Holmes, S. (2013). Phyloseq: An R package for reproducible interactive analysis and graphics of microbiome census data. *PloS One*, *8*(4), e61217. https://doi.org/10.1371/journal.pone.0061217

Michonneau, F., Brown, J. W., & Winter, D. J. (2016). Rotl: An r package to interact with the open tree of life data. *Methods in Ecology and Evolution*, *7*(12), 1476–1481. https://doi.org/10.1111/2041-210X.12593

Morgan, X. C., Segata, N., & Huttenhower, C. (2013). Biodiversity and functional genomics in the human microbiome. *Trends in Genetics*, *29*(1), 51–58. https://doi.org/10.1016/J.TIG.2012.09.005

Neher, R. A., Bedford, T., Daniels, R. S., Russell, C. A., & Shraiman, B. I. (2016). Prediction, dynamics, and visualization of antigenic phenotypes of seasonal influenza viruses. *Proceedings of the National Academy of Sciences*, *113*(12), E1701–E1709. https://doi.org/10.1073/pnas.1525578113

Page, R. D. M. (2002). Visualizing phylogenetic trees using TreeView. *Current Protocols in Bioinformatics*, *Chapter 6*, Unit 6.2. https://doi.org/10.1002/0471250953.bi0602s01

Paradis, E., Claude, J., & Strimmer, K. (2004). APE: Analyses of Phylogenetics and Evolution in R language. *Bioinformatics*, *20*(2), 289–290. https://doi.org/10.1093/bioinformatics/btg412

Pond, S. L. K., Frost, S. D. W., & Muse, S. V. (2005). HyPhy: Hypothesis testing using phylogenies. *Bioinformatics (Oxford, England)*, *21*(5), 676–679. https://doi.org/10.1093/bioinformatics/bti079

R Core Team. (2016). *R: A language and environment for statistical computing.* R Foundation for Statistical Computing. https://www.R-project.org/

Rannala, B., & Yang, Z. (1996). Probability distribution of molecular evolutionary trees: A new method of phylogenetic inference. *Journal of Molecular Evolution*, *43*(3), 304–311.

Retief, J. D. (2000). Phylogenetic analysis using PHYLIP. *Methods in Molecular Biology (Clifton, N.J.)*, *132*, 243–258.

Revell, L. J. (2012). Phytools: An R package for phylogenetic comparative biology (and other things). *Methods in Ecology and Evolution*, *3*(2), 217–223. https://doi.org/10.1111/j.2041-210X.2011.00169.x

Sanderson, M. J. (2003). r8s: Inferring absolute rates of molecular evolution and divergence times in the absence of a molecular clock. *Bioinformatics*, *19*(2), 301–302. https://doi.org/10.1093/bioinformatics/19.2.301

Schliep, K. P. (2011). Phangorn: Phylogenetic analysis in R. *Bioinformatics*, *27*(4), 592–593. https://doi.org/10.1093/bioinformatics/btq706

Schloss, P. D., Westcott, S. L., Ryabin, T., Hall, J. R., Hartmann, M., Hollister, E. B., Lesniewski, R. A., Oakley, B. B., Parks, D. H., Robinson, C. J., Sahl, J. W., Stres, B., Thallinger, G. G., Van Horn, D. J., & Weber, C. F. (2009). Introducing mothur: Open-source, platform-independent, community-supported software for describing and comparing microbial communities. *Applied and Environmental Microbiology*, *75*(23), 7537–7541. https://doi.org/10.1128/AEM.01541-09

Schmidt, H. A., Strimmer, K., Vingron, M., & Haeseler, A. von. (2002). TREE-PUZZLE: Maximum likelihood phylogenetic analysis using quartets and parallel computing. *Bioinformatics (Oxford, England)*, *18*(3), 502–504.

Schön, I., Shearn, R., Martens, K., Koenders, A., & Halse, S. (2015). Age and origin of Australian Bennelongia (Crustacea, Ostracoda). *Hydrobiologia*, *750*(1), 125–146. https://doi.org/10.1007/s10750-014-2159-z

Segata, N., Izard, J., Waldron, L., Gevers, D., Miropolsky, L., Garrett, W. S., & Huttenhower, C. (2011). Metagenomic biomarker discovery and explanation. *Genome Biology*, *12*(6), R60. https://doi.org/10.1186/gb-2011-12-6-r60

Shoemaker, J. S., & Fitch, W. M. (1989). Evidence from nuclear sequences that invariable sites should be considered when sequence divergence is calculated. *Molecular Biology and Evolution*, *6*(3), 270–289.

Stamatakis, A. (2014). RAxML version 8: A tool for phylogenetic analysis and post-analysis of large phylogenies. *Bioinformatics*, *30*(9), 1312–1313. https://doi.org/10.1093/bioinformatics/btu033

Venkatesh, D., Poen, M. J., Bestebroer, T. M., Scheuer, R. D., Vuong, O., Chkhaidze, M., Machablishvili, A., Mamuchadze, J., Ninua, L., Fedorova, N. B., Halpin, R. A., Lin, X., Ransier, A., Stockwell, T. B., Wentworth, D. E., Kriti, D., Dutta, J., Bakel, H. van, Puranik, A., … Lewis, N. S. (2018). Avian influenza viruses in wild birds: Virus evolution in a multihost ecosystem. *Journal of Virology*, *92*(15). https://doi.org/10.1128/JVI.00433-18

Vos, R. A., Balhoff, J. P., Caravas, J. A., Holder, M. T., Lapp, H., Maddison, W. P., Midford, P. E., Priyam, A., Sukumaran, J., Xia, X., & Stoltzfus, A. (2012). NeXML: Rich, extensible, and verifiable representation of comparative data and metadata. *Systematic Biology*, *61*(4), 675–689. https://doi.org/10.1093/sysbio/sys025

Wang, L.-G., Lam, T. T.-Y., Xu, S., Dai, Z., Zhou, L., Feng, T., Guo, P., Dunn, C. W., Jones, B. R., Bradley, T., Zhu, H., Guan, Y., Jiang, Y., & Yu, G. (2020). Treeio: An r package for phylogenetic tree input and output with richly annotated and associated data. *Molecular Biology and Evolution*, *37*(2), 599–603. https://doi.org/10.1093/molbev/msz240

Wickham, H. (2016). *ggplot2: Elegant graphics for data analysis*. Springer. http://ggplot2.org

Wilgenbusch, J. C., & Swofford, D. (2003). Inferring evolutionary trees with PAUP*. *Current Protocols in Bioinformatics*, *Chapter 6*, Unit 6.4. https://doi.org/10.1002/0471250953.bi0604s00

Wilkinson, L., Wills, D., Rope, D., Norton, A., & Dubbs, R. (2005). *The Grammar of Graphics* (2nd edition). Springer.

Wong, V. K., Baker, S., Pickard, D. J., Parkhill, J., Page, A. J., Feasey, N. A., Kingsley, R. A., Thomson, N. R., Keane, J. A., Weill, F.-X., Edwards, D. J., Hawkey, J., Harris, S. R., Mather, A. E., Cain, A. K., Hadfield, J., Hart, P. J., Thieu, N. T. V., Klemm, E. J., … Dougan, G. (2015). Phylogeographical analysis of the dominant multidrug-resistant H58 clade of salmonella typhi identifies inter- and intracontinental transmission events [Journal Article]. *Nature Genetics*, *47*(6), 632–639. https://doi.org/10.1038/ng.3281

Wu, N., Yang, X., Zhang, R., Li, J., Xiao, X., Hu, Y., Chen, Y., Yang, F., Lu, N., Wang, Z.others. (2013). Dysbiosis signature of fecal microbiota in colorectal cancer patients. *Microbial Ecology, 66*(2), 462–470. https://doi.org/10.1007/s00248-013-0245-9

Xu, S., Chen, M., Feng, T., Li, Z., Lang, Z., & Yu, G. (2021). Use ggbreak to effectively utilize plotting space to deal with large datasets and outliers. *Frontiers in Genetics, 12*, 774846. https://doi.org/10.3389/fgene.2021.774846

Xu, S., Dai, Z., Guo, P., Fu, X., Liu, S., Zhou, L., Tang, W., Feng, T., Chen, M., Zhan, L., Wu, T., Hu, E., Jiang, Y., Bo, X., & Yu, G. (2021). ggtreeExtra: Compact Visualization of Richly Annotated Phylogenetic Data. *Molecular Biology and Evolution, 38*(9), 4039–4042. https://doi.org/10.1093/molbev/msab166

Yang, Z. (2007). PAML 4: Phylogenetic Analysis by Maximum Likelihood. *Molecular Biology and Evolution, 24*(8), 1586–1591. https://doi.org/10.1093/molbev/msm088

Yang, Z. (1994). Maximum likelihood phylogenetic estimation from DNA sequences with variable rates over sites: Approximate methods. *Journal of Molecular Evolution, 39*(3), 306–314.

Yu, G. (2020). Using ggtree to visualize data on tree-like structures. *Current Protocols in Bioinformatics, 69*(1), e96. https://doi.org/10.1002/cpbi.96

Yu, G., & He, Q.-Y. (2016). ReactomePA: An r/bioconductor package for reactome pathway analysis and visualization. *Molecular BioSystems, 12*(2), 477–479. https://doi.org/10.1039/C5MB00663E

Yu, G., Lam, T. T.-Y., Zhu, H., & Guan, Y. (2018). Two methods for mapping and visualizing associated data on phylogeny using ggtree. *Molecular Biology and Evolution, 35*(12), 3041–3043. https://doi.org/10.1093/molbev/msy194

Yu, G., Smith, D. K., Zhu, H., Guan, Y., & Lam, T. T.-Y. (2017). Ggtree: An r package for visualization and annotation of phylogenetic trees with their covariates and other associated data. *Methods in Ecology and Evolution, 8*(1), 28–36. https://doi.org/10.1111/2041-210X.12628

Yu, G., Wang, L.-G., Han, Y., & He, Q.-Y. (2012). clusterProfiler: An r package for comparing biological themes among gene clusters. *OMICS: A Journal of Integrative Biology, 16*(5), 284–287. https://doi.org/10.1089/omi.2011.0118

Yu, G., Wang, L.-G., & He, Q.-Y. (2015). ChIPseeker: An R/Bioconductor package for ChIP peak annotation, comparison and visualization. *Bioinformatics, 31*(14), 2382–2383. https://doi.org/10.1093/bioinformatics/btv145

Yu, G., Wang, L.-G., Yan, G.-R., & He, Q.-Y. (2015). DOSE: An r/bioconductor package for disease ontology semantic and enrichment analysis. *Bioinformatics, 31*(4), 608–609. https://doi.org/10.1093/bioinformatics/btu684

Zmasek, C. M., & Eddy, S. R. (2001). ATV: Display and manipulation of annotated phylogenetic trees. *Bioinformatics, 17*(4), 383–384. https://doi.org/10.1093/bioinformatics/17.4.383

Index